# A Guide to the Russian Federation Air Force Museum at Monino

# A Guide to the Russian Federation
# Air Force Museum at Monino

B. Korolkov & V. Kazashvili

**Schiffer Military/Aviation History**
Atglen, PA

Technical editing by M. Gallhi and George Mellinger. The publisher wishes to thank Mr. George Mellinger and Ms. Olga Abramova for their assistance with this project.

Book Design by Ian Robertson.

Printed in China.
ISBN: 0-7643-0076-8

We are interested in hearing from authors with book ideas on related topics.

Published by Schiffer Publishing Ltd.
77 Lower Valley Road
Atglen, PA  19310
Phone: (610) 593-1777
FAX: (610) 593-2002
Please write for a free catalog.
This book may be purchased from the publisher.
Please include $2.95 postage.
Try your bookstore first.

# CONTENTS

# Introduction

Some 38 kilometers from Moscow in a deep pine forest along the straight-line highway connecting the capital of the Russian Federation with the city of Nizhni Novgorod is the Russian Federation Air Force Museum at Monino—a truly unique collection.

The museum was founded on a rich collection of full-scale aircraft exhibits, as well as helicopters, aircraft engines, armament, and search-and-rescue equipment, and reflects the history of Russian and Soviet aviation development from 1909 to the present. Besides full-sized exhibits, the museum houses many scale models and other exhibits containing unique documents and rare archival material—over 37,000 items.

The Air Force Museum is much broader in the aviation sense, as it contains displays relating to the USSR and Russia Air Force, as well as civil aviation and full-scale passenger airliners.

The airfield near the museum buildings was founded in 1928 in a former forest meadow and its adjacent marshes, and occupied a rather vast territory. At first the area was not wide enough for construction of the airfield, which is why hundreds of pines, stumps, and bushes had to be removed in order to make the surface even for laying taxiways. Two runways (each 850 meters long) were designed for the take-offs and landings of new 4-engine TB-3 heavy-bombers. Huge hangars, offices, dwellings, and a one-way railway-line were also built at the same time. In 1932 the first TB-3s started landing on the runways. In 1940 the Air Force Academy was also stationed there.

During World War II (the Great Patriotic War) the Monino airfield served as a home base for regular Air Force units, in particular the Il-4 long-range bomber units.

Post-war aircraft turned out to be too large to be handled by such a small airfield, and, since it was impossible to lengthen the runways, the airfield was closed for flight operations on April 4, 1956.

Early in 1958, Marshal of Aviation and Air Force Academy Chief S. Krasovsky issued an order forming a group of specialists that were assigned the mission of reconstructing and restoring the former Academy repair shops for their future employment as museum exposition buildings. Major-General (Ret.) Michael Shishkin was appointed chief of the group. The half-ruined wooden hangar, at 4,500 square meters, along with the adjoining buildings, were the major sites to be reconstructed. Later on the hangar was converted into the main exposition hall housing the Great Patriotic War aircraft. Two additional unheated hangars with the same area of 4,500 square meters were later attached to the museum.

Officially, the Air Force Museum was established on the Order of Commander-in Chief and Marshal of Aviation K. Vershinin on November 28, 1958. Order 209, "On the establishment of the Museum-Exhibition of the Air Force aviation equipment at the Red Banner Air Force Academy," read:

"Soviet military educational institutions, research institutes, repair shops, Air Force units, shooting ranges, and many other offices have acquired considerable quantities of aviation equipment, models, documents, relics, and various air-history artifacts. In due course they somehow get lost or destroyed. Therefore it is urgent to collect all remaining models of aircraft, engines, weapons, special equipment, airborne assets, documents, and relics on the history of the development and combat employment of our aviation.
Hereby I order:
To establish a Museum-Exhibition of Air Force aviation at the base of the Red Banner Air Force Academy. The exposition of the museum must reflect the development and combat employment of home aviation."

The museum staff was primarily composed of a director, a deputy-director, three researchers, and three hall-keepers.

On February 23, 1960, the museum was opened to the public. It was later converted into "The USSR Air Force Museum." In 1992 it was renamed "The Russian Federation Air Force Museum", and remains so today.

The main difficulty facing the staff of the museum since its foundation has been the search for aviation equipment and especially aircraft, particularly those older models which were put out of service long ago.

Unfortunately, very little care was taken with older aircraft in our country, as well as in most countries of the world, judging by existing aviation museums. The Central Aviation and Cosmonautics Association, for instance, contributed greatly to the call for aircraft restoration. Some aircraft downed during World War II were found in fields, woods, or marshes, and rebuilt. However, a few aircraft, essential for the museum exposition to be as complete as possible, are represented only as full-size mock-ups or models. The museum exhibits are mainly original aircraft that at one time or another were operational.

At present the Air Force Museum displays over 108 aircraft, 18 propeller-driven flying vehicles, 14 gliders, 120 engines, 2512 rockets, missiles, and aviation equipment models, and three ejection seats.

Since the earliest days of its existence the museum restoration shop, as well as most of the leading design bureaus—OKBs, named after A. Tupolev. S. Ilyushin, O. Antonov, A. Mikoyan, and others—have made a valuable contribution to museum reconstruction and making exhibits suitable for display. The process of adding new exhibits to the museum is now in full swing.

In 1958 Major K. Danilin was appointed the first Air Force Museum Director. From 1961-1972 this post was occupied by Major-General of Aviation A. Shcherbakov; from 1972-1982 by Lieutenant-General of Aviation S. Fyodorov; and from 1989-1992 by Lieutenant-General of Aviation V. Gorbachev. At present the museum is headed by Colonel V. N. Tolkov.

The book you are holding in your hands is devoted mostly to full-sized exhibits and, partly, to a small number of models built in the generally accepted scales.

The vast exposition and the thematic hall's materials (the documents of the hall devoted to Soviet aviation activities during the Great Patriotic War, for instance), as well as the rich scientific archives containing over 1,500,000 original historical documents, are not included in the book.

Most of the flying craft—mass production types in particular—were subjected to serious modifications followed by significant changes in their flying characteristics. So, in this book one can find the characteristics of either the latest modification or greatest mass production type.

Since its establishment over 3 million visitors have been to the museum. The tour around the museum is highly informative due to the skilled guide staff composed mainly of former pilots, test pilots, navigators, and military aviation engineers. Here is what Indian Air Force Brigade-General Jasdkith Sing, director of the Defense Research Institute, wrote to the authors: "A visit to your museum was one of the most impressive events of my trip. For me, a military pilot who has spent over 30 years flying fighters, this visit was like a home-coming. Having visited nearly all of the world's aviation museums, I can't help admiring the people who have made your museum the best—and quite a unique one."

If there were discrepancies in some sources on the official designations of certain aircraft, the preference was given to official publication information: technical firm descriptions, flying test reports, etc.

In the museum record book there are also the following impressions of Doctor of Medicine Natalya Koroleva, daughter of space systems designer S.P. Korolev: "We have visited the Air Force Museum with great interest. No doubt, this is a unique aviation museum in size, form, and content. While visiting one feels proud of our motherland and its splendid designing spirit. I'd like to express deep gratitude to the founders and keepers of the museum, its splendid guides, great professionals, and the entire museum staff."

Mr. Paul Delphy, an Irish journalist, is of the same opinion: "I've been to 30 aviation museums throughout the world. I knew there existed a museum in Monino. I had expected to see a lot there, yet I was quite shocked with what I saw there. It's the most attractive museum of all, containing the richest aircraft collection. It's just fantastic."

It goes without saying that flipping through this book cannot substitute for a visit to the museum, just as a single day visit to the museum is not enough time for a careful examination of its collections.

Nonetheless, the authors hope this book will give readers a general idea of the museum, arouse in them a desire to perhaps one day visit, and also serve as a reliable source of information on the exhibits reflecting the Russian and Soviet aviation history.

The authors are grateful to Yu. Zassypkin, V. Ilyushin, D. Kantor, L. Selyakov, L. Egenburg, L. Davidov, and S. Federov for their contribution.

# 1

## Aviation in the Late XIX-Early XX Centuries

### "Letatlin"

It has been a long time since the air and space first attracted mankind. In order to fulfill a strong desire to fly, people naturally tried to follow the example of a bird soaring, an example given by nature itself, which is why even up to the present many attempts have been made to design various ornithopters (heavier-than-air craft with flapping wings).

Leonardo da Vinci firmly believed in the possibility of "flapping wing" flight, and he himself designed a number of flying machines with such wings.

From 1929-1933 Vladimir Tatlin, a pencil artist and painter, a talented scene-painter, architect, and inventor, worked at designing a similar flying machine. He managed to create "Letatlin", a most significant flying craft from the point of view of form. Its moving parts are mounted on ball-bearings, and it is encased in a silky skin. The "Letatlin" is characterized by: weight - 32 Kg; wing span - 12 square meters; load - 8kg/square meter.

Though the flight of his apparatus was destined to fail (it was doomed to failure, judging by the subsequent calculations), this machine is of great historical significance as it is one out of quite a number of attempts to create an ornithopter. The "Letatlin" was exhibited at the Pushkin Museum of Fine Arts, at the Central House of Literary Men, and in the USA. It has been displayed at the Air Force Museum since 1958.

### A. Mozhaisky "Flying Vehicle" (Model)

In 1882 a Russian naval officer began constructing the first full-scale aircraft designed for a manned flight. The aircraft possessed all the essential elements which are present in today's aircraft design: fixed wings, a fuselage (hull-type), a tail unit, a power plant, and a gear (wheeled bogie undercarriage).

Unfortunately, the aircraft construction was carried out too slowly as the funds granted on it from the Tsar's treasury were given most unwillingly and with great delays. Therefore, the designer had to spend a lot of his own savings on the work.

Two 10 HP and 20 HP steam engines were installed in the machine. The first airscrew was set into motion by the 10 HP steam engine, while the other two small airscrews were brought into a rotary motion by means of the 20 HP steam engine. Both steam engines were produced by the "Arbaker" firm, England, to the special order of A. Mozhaisky.

The fuselage had wooden edges with cloth covering. The rectangular wings attached to the fuselage were a little bent slightly upwards. The apparatus was mounted on a wheeled strut and weighed 1077 kg.

Though the precise aircraft drawings of A. Mozhaisky were not found, a careful and detailed comparison of different papers done by the historians made it possible to form a true notion of its original form.

Examining the archive documents one can come to the conclusion that in 1883-1885, A. Mozhaisky was occupied with the aircraft improvement in ground tests, and during the second part of July 1885 he attempted to conduct its flying tests on the military airfield near Krasnoye Selo. The flying machine, unluckily, bent awry, and its supporting surfaces were broken. The damage was not great, but

A. Mozhaisky could not afford its reconstruction, having run out of money.

With the creation of the first aircraft the mono-plane type with a steam-engine, A. Mozhaisky, an outstanding scientist and engineer, surpassed all the ideas of science and technology of his time.

The Air Force Museum has an A. Mozhaisky model on display.

Further scavenging and aerodynamic calculations proved that Mozhaisky's craft was incapable of flight, as the power of the engine was considerably smaller (approximately three times) than that necessary for its horizontal flight, to say nothing of the take-off. Nevertheless, the importance of the development of the first life-sized flying apparatus, the scheme of which fully justified itself, was immense. The precedent it set was not ignored by future generations of aircraft designers in Russia and abroad.

# The Wright Brothers Aeroplane (Model)

The Air Force Museum exhibits an aircraft model by the Wright brothers, Wilbur and Orville, the American inventors and aircraft designers. There is also a full-scale 25 HP engine of their design on display at the museum.

The Wright brothers designed various types of gliders, and made over 1,000 flights in them. In the biplane-glider they fixed an internal combustion engine, 16 HP thrust, and on December 17, 1903, Orville Wright performed the world's first successful flight, which lasted twelve seconds (Kitty Hawk, North Carolina). Three more flights, lasting 13, 15, and 19 seconds, were made by him on the same day. The flying machine took off from the platform, using a catapult, and flew practically straight.

The Wright Brothers' aircraft possessed a rudder (two turning fins behind the wing); an elevator (a movable destabilizer in front of the wing); and a slight skew of the edges of the wings. The rudder and the skewed wings' edges were joined by a control-lever which provided a banked turn.

Two pusher propellers were fixed behind the wing cell and connected to the engine by a chain transmission, which was mounted in the middle of the lower wing. The skis served as a take-off and landing device. The aircraft weight was 415 kg; span - 12.25 meters; speed - 4-km/h.

In 1905 the Wrights achieved such great progress that their invention was talked about throughout the world. In 1908 they improved their plane and managed to make its flight with some passengers on board. Though the aircraft construction scheme wasn't widely spread, it went down in the world aviation history as the first heavier-than-air craft to perform an engine-powered manned flight.

# Bleriot Aircraft (Model)

The Air Force Museum exhibits include a model of the aircraft designed by French Pilot and designer Louis Bleriot. The 25 HP "Anzani" engine, which was installed in the aircraft, is demonstrated at the museum, too.

The monoplane of Bleriot, an original European design (1909), had a fuselage and a tractor propeller. It differed greatly from the Wrights' craft and was typical of piston-engine fliers in design. On July 25, 1909, Louis Bleriot made a flight via La Manche from France to England. It was his eleventh machine. Bleriot took off from Callais and landed in Dover, England, in 37 minutes. It was considered to be an outstanding flight of that time, the first one in the world performed over water.

Besides, Louis Bleriot was the first to create monoplanes (the scheme developed by Mozhaisky), which later on proved to be a prevailing type of aircraft for a long period of time.

## A. Ufimtsev Engine

Anatoly Ufimtsev, a designer and an inventor, developed two aircraft of the original "spheroplane" design (with a planned round wing). Neither of his machines, however, were ultimately capable of flight. Ufimtsev also designed four aircraft engines.
One of these engines, created by Ufimtsev for his "spheroplane - 2", is displayed at the Air Force Museum. Its designation is ADU-4. The ADU-4 is a 6-cylinder, radial, birotative, 2-contact air-cooled engine with swimming cylinders, 100 HP thrust, and a weight of 58 kg. Of note is the fact that a considerably small specific gravity of the engine testifies to the perfection of the design, especially at the time it was created (1909-1911).
The tsarist Russian aircraft industry didn't have mass produced engines of its own. Nevertheless, such experienced designers as Ufintsev, for instance, showed that the Russian technical innovators worked enthusiastically.

## "Grizodubov-1" Aircraft (Model)

Stefan Grizodubov was one of the first Russian aviators and designers. For a three-year period between 1909 and 1912, he constructed four aircraft and engines of his own design.
In 1909, S.V. Grizodubov started constructing the first aircraft. In addition, he designed a 4-cylinder air-cooled engine with 40 HP thrust to accompany it, both of which were built in 1910. All of the work, except the cylinder widening, was carried out by the designer himself in his own workshop. The engine was installed in three of Grizodubov's aircraft. Each plane was constructed by using some components of the previous aircraft model.
The Air Force Museum displays a "Grizodubov-1" aircraft model and its full-sized engine.

Within the following years, Grizodubov was occupied with the construction of the second and then the third craft, but the aircraft engines were not powerful enough for the take-off.
Taking that fact into consideration, Grizodubov stopped remodeling the existing designs and created quite a new aircraft, which resembled a "Bleriot-XI" and had the "Anzani" engine, with a thrust of 25 HP, installed in it. The new model turned out to be rather successful, and Grizodubov made a number of flights in it in 1912. But it was Grizodubov's daughter, Valentina, who made his name world famous. She was a famous pilot who performed a number of record long-distance flights. During the war, she commanded a bomber regiment.

## "Ilya Muromets" Aircraft (Full-Scale Mock-Up)

Constructing the world's largest multi-engine aircraft, Igor Ivanovich Sikorsky, an outstanding aircraft designer, had already become popular owing to a light one-seater he made. He started constructing a multi-engine aircraft in spite of the doubts expressed by many competent specialists of his time.
Still, even his first version could fly pretty well. The aircraft got the name of "Grand Baltic", and was later renamed "Russian Vityaz" (Span-70m, gross weight-over 4 tons).

The "Ilya Muromets", with the "Russian Vityaz" modification, represented a new stage in world aircraft construction. In fact, it served as a prototype for all multi-engine bombers. There appeared some "Ilya Muromets" modifications, which differed mainly in the structure of the power plant and had some other specific features.
For the first time in the world, it became standard engineering practice that the fuselage of the "Ilya Muromets" was made one-piece, with a tetrahedral section exceeding a man's height. In the front

fuselage there was a cabin (8.5 meters long; 1.6 meters wide; 2 meters high). Its weight varied from 4,800 to 6,100 kg, and the wing area was 200 square meters The span on the upper wing in different modifications varied from 24 to 34.5 meters, the lower wing from 17 to 27 meters. Only the upper wing was comprised of ailerons. The aircraft could carry up to 480 kg of bombload, which was originally dropped mechanically by means of a handle, and since 1916 with the help of an electrodropper first employed in such aircraft around the world.

There were four "Argus" engines in the aircraft, 100 HP each. It also had two "Salmon" engines, 200 HP each, installed internally, and two "Argus" 115 HP engines externally.

Sometime later the aircraft were equipped with 220 HP "Renault" engines.

Its first version was made in October 1913. After its production, the aircraft was employed for show flights. Sikorsky himself flew it. Its maximum speed gained was up to 130 kph.

The "Ilya Muromets" aircraft broke all the world records in range and carrying capacity at that time. A record in carrying up to 1,300 kg, with fifteen passengers and a dog on board, was set on February 12, 1914. That record was considered to be second-to-none till 1919. In June 1914 a flight from "Petersburg-Kiev" and back was performed.

In August 1914 the "Ilya Muromets" went into service with the Russian Army. It was the world's first specifically designed bomber. The "Ilya Muromets" heavy machines were organized into an "air ships squadron", and used as bombers and reconnaissance planes. The Air Force Museum displays an "Ilya Muromets" mock-up with the working engines made in Czechoslovakia. It was constructed life-sized to the order of the "Mosfilm" studio, especially for the shooting of the film "Poem on Wings." The mock-up can taxi and make runs.

The Air Force Museum got the mock-up in 1979. It has been demonstrated here since 1985, after the reconstruction.

As to Sikorsky, he emigrated from Russia in 1918 and later, working in the USA, he constructed a number of outstanding flying craft, large-scale hydroplanes (flying boats), and wide-spectrum helicopters.

During the First World War a major part of the Russian military air fleet was of foreign design, with most of the production abroad and mainly in France. The first flying machines employed in Russia were aircraft from the firm of "The Voisin brothers, Gabriel and Charles"

In wartime it was used as a reconnaissance plane. Its armament: one machine-gun and 45-50 kg bombload.

# "Voisin"

On January 15, 1915, the owner of the aircraft factory in St. Petersburg, S.S. Schetinin, signed a contract for the production of 50 "Voisin"-type aircraft. Since early 1916 the "Voisins" were produced by his plant.

In the Air Force Museum exposition there is a "Voisin" aircraft made in 1916. It is a biplane with a pusher propeller. A 150 HP "Salmson" engine is installed in it.

SPECS: Max speed was 105-110 km/h; landing speed - 70 km/h; it could climb to the altitude of 2,000 meters in 23 minutes; ceiling-over 3,000 meters; length 9.5 meters; wing area - 42 square meters. The "Voisin" aircraft was employed in shooting the films "The Elusive Avengers" and "There Served Two Friends."

# "Sopwith"

The Air Force museum exposition includes a model of the "Sopwith" aircraft produced on March 28, 1959. It was presented to the museum by the Central Aviation and Cosmonautics Club.

"Sopwith" is a tri-plane one-seater fighter, English design, made at the V. Lebedev aircraft factory in 1917. The plane had a "Clerpet" 130 HP engine, rotary type, made in England. Max Speed - 198 kph; weight - 642 kg; span - 8.1 meters; length 5.9 meters; height - 3m.

Since it was used in Russia mainly as a trainer, the aircraft was not armed. It was characterized by high maneuverability, according to the pilots' reports.

But, on the whole, a triplane scheme was not successful and was employed in only a few types of aircraft (Jupi, Waneman, Besobrasov, KOMTA, etc.). The triplane has not been used since 1920, since there are no advantages over a biplane scheme. A Soviet scientist, Professor V. Pishnov, pointed out its low pilot visability, as well as insufficient stiffness of the high three-wing cell and its overall bulkiness.

# "Farman IV"

This flying craft represents one of the flying machines designed by a French designer and pilot (such co-professions were rather typical for that time). Henri Farman designed the aircraft together with his brother Maurice Farman.

On March 8, 1910, a Russian pilot, Efimov, one of the first pilots in the country, made the first Russian show flight in this airplane for the spectators of the Odessa racecourse (no airfields existed at that time). Another popular Russian aviator, S. Utochkin, took part in the show flight operating the same aircraft. Such pilots as L. Matschkevich, E. Rudnev, and others started their flying career and soon became widely known for flying the "Farman-IV" machine. From 1909-1916 the "Farman-IV" biplane was the most widely used aircraft in Russia. Many aviators learned to fly by operating it. It was not practically used for military purposes. The aircraft was employed mostly for shows, as well as for pilots' training, although

its piloting qualities (for example, stability and controllability) were far from perfect.

SPECS: Upper wing span - 10.5 meters; length of the aircraft - 12.50 meters; weight - 5,400 kg; load mass - 180 kg; max speed - 65 kph; landing speed - 60 kph. It possesses a rotary type "Gnome" engine, 50 HP thrust.

The "Farman-IV" exhibit displayed at the museum was designed for a special order of the "Lenfilm" studio by the DOSAAF avia-sports-club in Leningrad. They made three flying machines. A special engine, which differs from the original, was mounted in the aircraft. It was the "Farman-IV" plane which was shot in the film "Aeronauts." 50 flights were performed in it. Captain Kuchma, chief of the Leningrad avia-sports-club, flew this aircraft, too. The aircraft model was delivered to the Air Force Museum on September 6, 1975, dismantled. It needed urgent repairs, which were carried out rather successfully at the museum repair shop.

# 2

## Aircraft of the 1920s-1930s

## The ANT-2

The ANT-2 was the first all-metal aircraft in our country, designed by A. Tupolev as an experiment in applying metal in the process of aircraft building. All the numerous aircraft models that followed were known to be all-metal, though in the beginning many specialists considered that trend in aircraft construction to be rather doubtful. There was a saying, "Russia is the country of woods. So, ignoring wood as an aircraft building material is most unreasonable."

It stands to reason that a plane like the ANT-2 might as well be made of wood, yet Tupolev and his colleagues thought otherwise. It was not the ANT-2 itself they needed. What they wanted for Russia was a prototype flying machine which would pave the way for a new trend in home aircraft construction, because up to that time, all-metal flying craft were produced in Germany only.

According to its design the ANT-2 is a cantilever-monoplane with the "Bristol-Lucifer" air-cooled engine, with 100 HP

SPECS:
weight - 836 kg; max speed - 170 km/h; ceiling - 3,000 meters; span - 10m; length - 7.5m; height - 2.13 meters.

The first flight of the ANT-2 was performed by test-pilot N.I. Petrov on May 26, 1924.

As was already mentioned; the designers made an experimental version of the ANT-2, yet it could operate with two or even three passengers on board rather well.

The Air Force Museum ANT-2 exhibit is an original, the only constructed one in existence.

It was delivered to the museum from the Central Aviation and Cosmonautics Club on October 11, 1960.

# The ANT-4 (TB-I) Model

ANT means Andrei Nikolaevich Tupolev. This abbreviation was adopted before the late 1930s. Later on a new system of designation according to the first letters of the chief designer's surname was introduced.

TB-I, the first heavy bomber, was the designation which existed in the Air Force, and reflected the combat function of the flying machine.

The Soviet Air Force needed development of a new combat aircraft fleet. So, one of the English aircraft building firms was to carry out the Soviet order on the construction of a heavy bomber for the USSR Air Force.

The Tupolev Design Office, TsAGI (the Central Aerohydrodynamic Institute), was not charged to do it at first, as by that time the TsAGI had only produced models for one light

ТБ-I с 2 моторами БМВ-у1
Тяжелый бомбардировщик.

plane (ANT-1), one experimental aircraft (ANT-2), and the only combat machine, reconnaissance plane ANT-3 (R-3), which was put into mass production.

But the English firm's production prices were too high. They requested 2 million dollars for the construction term of 2 years. Our country could not afford it, which is why we had to refuse English assistance and entrust the TsAGI with the mission. On November 11, 1924, they started designing and constructing the TB-I (ANT-4), which was to be powered by two "Napir-Lion" engines with 450 HP thrust each. Though the aircraft construction was carried out in quite an unsuitable building, the nine-month term of its construction was strictly observed.

On August 11, 1925, the aircraft was delivered to the Central airfield where it was assembled, and on November 26, 1925, test-pilot Tomashevsky made its first flight.

The TB-I was mass produced within a number of years and was constantly modified. There were over 200 aircraft, all of different versions. Its take off weight varied from 6,200 to 7,800 kg. (in different modifications); span 28.7 meters; wing area - 120 square meters; length - 18.0 meters. Its armament included 4-6 guns and from 1,000 to 3,000 kg of bombs. The crew consisted of 6 men. The max speed was 187-205 km/h. The "Napir-Lion" engines were, at the primary stage, replaced by the "VM-6", and then by the "M-17" engines, with a thrust of 500-715 HP The major part of this aircraft fleet was equipped with the "M-17" engines.

The TB-I served as the basis for the development of the Soviet heavy bomber aviation. It was also the first aircraft in the world to utilize heavy cantilever monoplane construction, with the engine mounted directly on the wing.

In 1929 the crew, consisting of pilots S. Shestakov and F. Bolotov, navigator B. Sterligov, and flight mechanic D. Fufaev, made the ANT-4 "Strana Sovjetov" flight (with stops) from Moscow to New York via Siberia, the Far East, and San Francisco, covering the distance of over 21,000 km in 142 flying hours. The flight had a great resonance among the USA public and proved to be a real test of the aircraft crew and equipment.

In March 1934, the TB-I crew, headed by pilot A. Lyapidevsky, was the first to land on the iceflow in the Sea of Chukotsk (Chuckchee Sea) and to come to the rescue of the people of the shipwrecked steamship "Chelyuskin." The aircraft crew rescued the women and the children and airlifted them to the mainland, thus beginning the rescue operation which was known to be carried out rather succcssfully.

The TB-I design and construction was the basis for the development of both its smaller (the ANT-7 reconnaissance plane R-6) and larger version (the ANT-6 heavy bomber -TB-3). This will be discussed further on.

The Air Force Museum exhibits the TB-I aircraft model (the scale is 1:20) created by the group of the aircraft model builders under the guidance of Lieutenant-Colonel G. Pritugin and handed over to the Museum on December 20, 1986.

# The ANT-6 (TB-3) Aircraft

The ANT-6 aircraft, constructed by A. Tupolev Design Office, was the world's first four-engine bomber—a monoplane with a cantilever wing.

The first flight of its experimental version, equipped with four Curtiss "Conqueror" engines, 600 HP thrust, took place on December 22, 1930, piloted by test-pilot Gromov. Later those engines were replaced by the M-17 (500-715 HP thrust), M-34 (750-830 HP thrust) engines, and since 1936 by AM-34RN engines (840-970 HP thrust)

The TB-3 (ANT-6) was the main heavy bomber of the Soviet Air Force. In all, the aircraft industry produced more than 800 such ships.

In 1937 an expedition headed by academic O. Shmidt landed on the iceflow in the North Pole, bringing along the first group of winterers, which marked the beginning of the polar investigations. The landing operation was performed by four ANT-6 aircraft, specially equipped for the flights in the Arctic, piloted by the crews of Vodopyanov, V. Molokov, A. Alexeev, and I. Mazuruk.

TB-3 combat machines were engaged in military conflicts in the Lake Khasan area and others. At the outbreak of the Great Patriotic War, the TB-3 had already become obsolete. However, it was still used in combat operations as an amphibian craft. SPECS: aircraft length- 25.18 meters; span 41.62 meters; wing area - 234.5 sq. meters; weight (in different modifications) - from 17,000 to 26,000 kg; max. speed (with the AM-34RN engines) - 288 kph; ceiling - 7,740 meters; radius of operation - 1,000 km; bomb load - up to 4,000 kg.

The Air Force Museum has a wooden model of the TB-3 aircraft on display, scale: 1:25. The museum got the model from the Central Aviation and Cosmonautics Club on December 28, 1959.

# The ANT-25 (RD)

The ANT-25 was specifically designed for long-range record flights, which is why it is also known under the name of RD, which means record in distance. The aircraft was designed and constructed by the team of designers headed by P. Sukhoi, under A. Tupolev's supervision. Taking into account the aircraft function, special attention was paid to its range, among such characteristics as speed, carrying capacity, and altitude. For the same reason, the aircraft had a peculiar outer appearance: the span of the cantilever wing, for instance, was unusually big for those years at 34 meters, and the length of the wing was far from being normal (13.1 m). The fuel tanks were the essential part of the wing configuration. The landing gear was semi-retractable.

Its first version was equipped with the M-34 engine, the second one with the M-34R (gear-driven) engine, thrust 800 HP It possessed the take-off weight of 11,500 kg.

On June 22, 1933, test-pilot M. Gromov made the first flight of the ANT-25.

The installation of a gear-driven engine on the ANT-25 increased its range considerably, though it was not sufficient enough to break the long-distance record of the French pilots Codos and Rossi (9100 km). This problem was solved by covering the corrugated skin of the wing with smooth fabric, which reduced the aerodynamic resistance to a great extent, thus increasing the range by more than 20 percent, and thereby causing the subsequent rejection of corrugated skin in the aircraft construction practice.

In the summer of 1934 the ANT-25, piloted by the crew of Gromov, A. Filin, and I. Spirin covered a distance of 12,411 km, setting the world record in range while flying on route. However, the record wasn't officially recognized for some formal reason: the USSR did not belong to the International Aviation Federation (FAI) at that time.

In August 1935 the crew, consisting of pilots S. Levanevsky and G. Baidukov, and navigator V. Levchenko, tried to perform a non-stop flight from Moscow to America via the North Pole. Unfortunately, it was not a success, because in the Barents Sea area the fuel poured out of the fuel tank drain. The crew commander found the breakage serious enough to discontinue the flight, and they flew back.

The next ANT-25 starts proved to be more effective.

On July 1936 the ANT-25, manned by V. Chkalov, G. Baidukov, and A. Belyakov, took off from the Shchelkovo airfield, located near Moscow, and in 56 hours 20 minutes landed on Udd Island, near the mouth of the Far East river Amur, having covered a distance of 9,374 km.

It was during that stop flight that the Chkalov's crew became adamant in their decision to fly only the given (ANT-25) machine to America via the North Pole.

At dawn, on June 18, 1937, the crew was permitted to take off, and it set its course to the North Pole. They managed to cross the North Pole and reach America in 63 hours, 16 minutes, landing on a military airfield not far from Vancouver, having covered a distance of 9,130 km (straight - 8,504 km)

Three weeks later another crew of pilots, Gromov, A. Jumashev and S. Danilin, performed the ANT-25 non-stop to America via the North Pole. At dawn on July 15, 1937, the aircraft landed in the vicinity of San Jacento, California. They covered a distance of 10,148 km in 62 hours 17 min. That was a real world record in distance flights.

The ANT-25 was put into series production in small number, 20 aircraft altogether, which, however, were of no practical use.

The Air Force Museum exhibits the ANT-25 technological copy made of the aircraft model which had performed the Moscow-San Jacento non-stop, piloted by the Gromov crew.

The "Chkalov" version of the ANT-25 is on display at the memorial museum which is located in the town of Vasilevo (Chkalov), Mizhegorodskaya Region.

# The ANT-40 (SB)

The fortieth aircraft made by A. Tupelov's Office had a military designation of SB- speedy bomber. Over the course of time it became clear that designating an aircraft with some of the flying characteristics of the machine was a risky proposition, as aviation developed rapidly and the models designated like that were replaced by new ones.

Nevertheless, the SB proved to be a remarkable aircraft for its time by its speed, and marked a great rise in aircraft performance. Moreover, for the first years of its existence it was hardly surpassed by any world fighter at that time.

The SB was characterized by a smooth configuration, smooth covering, rational wing-fuselage connection, fully retractable landing gear, and a great wing area load (up to 125 kg/sq.m.). The advanced aerodynamics, in combination with the powerful, high-altitude engines, provided high aircraft performance (the engines maintained the power at a definite altitude with the help of the pumps installed in them).

A. Tupolev charged A. Arkhangelsky's team with the task of constructing a new high-speed front-line bomber.

The first SB experimental version was powered by two "Wright-Cyclone" air-cooled engines, developing 730 HP each. Test-pilot K. Popov made its first flight on October 7, 1934.

Test-pilot N. Zhurov performed the first flight of its second experimental version, equipped with the "Kispano-Suiza"-12U engines, 780 HP thrust each, on December 30 that same year.

The maximum speed of the SB was 423 km/h. Later, it was increased to 450 km/h at an altitude of 4,000 m due to the replacement of the engines by more powerful M-100A (860 h.p) and then M-203 (960 HP) engines.

It is noteworthy that the SB aircraft turned out to be among a rare class in the world aircraft building practice as its flying performance appeared to be higher than it's original design specifications (two decades later the same firm happened to witness the similar situation with the Tu-16 jet bomber).

However, getting into the sphere of high speeds not developed before created an unpleasant surprise: while flying at the maximum speed, the SB experimental version had serious vibration problems of the flutter type. Test-pilot K. Minder managed to reduce speed, stop the vibrations, and land the plane. Weight-balancing of the ailerons helped to eradicate the defect completely. Since then that technique was widely used to avoid flutter.

The two-engine three-seater, accommodating a crew of a pilot, navigator, and a radio operator, was a mid-wing monoplane bomber which possessed the following performance (in mass production):

span - 20.33 meters; wing area - 56.7 square meters; length - 12.27 meters; normal take-off weight - 5,700 kg; overload take-off weight - over 6,500 kg; bomb-load - up to 1,000 kg; armament - 3-4 guns.

The aircraft industry produced a total of 6,656 SB aircraft.

It was employed with great success in the combat operations in Spain, the Far East, and in Mongolia, as well as in the military conflicts with Finland and even in the course of the Great Patriotic War (though aging by that time). Additionally, it was flown by civil aviation under the Il-40 designation.

The Air Force Museum possesses an SB model with an interesting history: the crew had to make a forced landing on the bank of the Ushkum River in the area of the Vitemskoye plateau, Transbaikalia, under instrument flight rules in the winter of 1939.

In 1979 the aircraft was transported to the museum by a Mi-8 helicopter and the AN-22 aircraft. The remaining parts were transferred to the "Opyt" Moscow machine-building plant (named after A. Tupolev) for reconstruction.

The aircraft repairs were carried out, and on August 15, 1982, General Designer Deputy A. Kondakov presented the SB aircraft to the Air Force Museum in a solemn ceremony. A. Arkchangelsky's wife took part in the grand occasion.

# The DB-3 (Il-4)

The DB-3 Aircraft, or "long range bomber the third", was created by the brigade of designers headed by S. Ilyushin in the Central Design Office (TsKB). That is why the designations of the first experimental models derived from the abbreviation TsKB.

In 1933 the TsKB was assigned the mission of designing a bomber with great radius of operation, bomb load, and altitude. All the existing machines of those days possessed great range but were supposed to be relatively slow-operating, with a great wing-span and a high-aspect-ratio wing. S. Ilyushin was the first to design an aircraft possessing high speed, long range, and high altitude, so the mission was carried out perfectly.

In the summer of 1935 test-pilot V. Kokkinaki first took the TsKB-26 into air, and in March the TsKB-30 flight took place. Those two aircraft differed in the following points: the first aircraft was a complete type construction (metal-wood), while the second one was all-metal. The latter one, designated the DB-3, entered service.

The first DB-3 machines were equipped with the M-85 engines made by license and based on the French "Gnome-Rom" Mistral-Major K-14 engine, 800 HP thrust. Ilyushin's next bombers were powered by the M-86, M-87, M-88 engines, in order to improve that aircraft family. The M-88 engine had the take-off power of 1,100 HP km/h.

The DB-3 was repeatedly modernized during the years of its mass production. The DB-3F version was renamed into the Il-4. It had the following performance: maximum speed - 445 km/h; ceiling - over 10,000 meters; take-off weight - up to 10,000 kg; normal bombload - 1,000 kg, maximum bombload - 3,000 kg; range - over 4,000 meters; wing span - 21.44 meters; wing area - 66.7 sq. meters; aircraft length - 14.8 meters.

On June 27-28, 1938, the serial DB-3b, piloted by test-pilot V. Kokkinaki and navigator A. Bryandinsky, made a non-stop flight from Moscow to the Far East, covering a distance of 758 km, at a medium speed of 307 km/h.

In April 1939 another aircraft of this series named "Moscow", piloted by V. Kokkinaki and meters. Gordienko performed a non-stop from Moscow to the USA (Moscow Isle), at a medium speed of 348 km/h, having covered a distance of nearly 800 km.

The DB-3 aircraft set a number of world records in height (with different load) and speed.

The DB-3 long-range bomber (Il-4) was mass produced. In all, 6,734 DB aircraft of all modifications were produced.

S. Ilyushin's long-range bombers were engaged in the Great Patriotic War and constituted a major part of the long-range aviation fleet. The first raid by Soviet aviation on Berlin was performed by our pilots in the DB-3, taking off from Kodu Air base (Sarremaa Isle, Estonia) in the pre-dawn hours of August 7-8, 1941.

The DB-3 aircraft displayed at the Air Force Museum was found in the Ussuriisk Taiga, some 120 km away from Komsomolsk-on-Amur. In September 1988 it was airlifted to Irkutsk by Il-76. After its reconstruction at the Irkutsk aircraft factory it was delivered to the Air Force Museum by the AN-22 transport plane (the crew of Russin) on December 21, 1989. On December 22-23 the Irkutsk aircraft factory representatives, with V. Zelenkov at the head, assembled the aircraft model.

# The U-2 (Po-2)

Designed in N. Polikarpov's creative collective in 1927, the U-2 aircraft ("trainer the second") was renamed the Po-2 ("Polikarpov the second") after the death of the designer in 1944. Because of its main function as a primary trainer, the aircraft was produced in a number of modifications, including as an ambulance-plane and agricultural plane (for aeropollinating fields). During the war it was even employed as a night light bomber, which turned out to be rather effective.

On January 7, 1928, test-pilot meters. Gromov made the first U-2 flight, thus marking the beginning of more than thirty-five years of service for this aircraft. Its excellent performance was revealed in the course of the flying tests. The U-2 was a reliable and cheap machine and easy to handle, and thus meeting its purpose completely. The aircraft was mass produced in great numbers and in different modifications at the serial plants until 1953, and in Aeroflot up until 1959.

In the pre-war and war years thousands of cadets from the flying schools, as well as boys and girls from the flying clubs, got a "start to the sky" operating the U-2 (since 1944 the Po-2). The majority of the pilots burdened with all the hardships of air combats during the Great Patriotic War were sure to get the first basic lessons in their career as pilots flying the U-2.

The U-2 is a two-seater biplane, wooden construction (veneer percale), equipped with the 100-110 HP M-11 engine, A. Shvetsov's design.

Its main dimensions are the following: span - 11.4 meters; length - 8.2 meters; wing area - 33.15 square meters; wight - 1,000-1350 kg; the flying characteristics: max speed - up to 145 kph; range -

470 km; crew - two men; armament - in the light bomber version: bomb load - 200-300 kg; 7.62 mm ShKAS gun.

The total number of the manufactured U-2/Po-2 aircraft exceeds 40,000.

The 46th Guards Women Air Regiment, which made over 25,000 sorties under the command of E. Bershanskaya, was the most famous among more than a hundred air units operating such aircraft.

The Po-2 aircraft, exhibited at the Air Force Museum, was produced at the aircraft factory on August 2, 1945, and delivered to the museum on November 21, 1958. The exhibit is in good exposition form. It was shot in the film "Night Witches in the Sky."

# The R-5

In the late 1920s-early 1930s a rather popular type of combat aircraft in many countries of the world was a two-seater, one-engine, reconnaissance-light bomber, normally a biplane or a sesquiplane. The Soviet Air Force's R-1 and R-3 were just such machines, though aging by the late 1920s.

The development of the R-5 reconnaissance plane turned out to be a remarkable event in the N. Polokarpov Design Office, as well as in the history of Soviet aviation. It was a sequiplane with an upper wing span of 15.5 meters; wing area - 50 sq. meters; wood construction with material wing and tail unit skin and veneer fuselage skin; 680 HP M-17 engine; weight (normal - 2,6000 kg; max 2,350 kg). The aircraft possessed excellent flying characteristics: max speed of up to 230 kph; ceiling - 6,400m; bomb load - 300-500 kg; range - 1,000 km.

It was test-pilot meters. Gromov who performed the first flight of this machine and its primary production tests in the beginning of 1929.

A number of test flights to Turkey, Afghanistan, and Iran were made in the R-5. In the fall of 1930 the aircraft won first place in the international competition in Teheran, having met all the requirements of the contest and leaving behind all the other competitors, some solid aircraft of the famous POTEZ firm among them.

In 1934, during the rescue operation, the R-5 aircraft piloted by V. Molokov and meters. Vodopyanov and N. Kamanin carried 83 men out of 104 crew members of the "Chelyuskin" steamship from the iceflows to the mainland.

There appeared attack (ShR-5), civil (P-5), and other versions of this aircraft which proved to be rather successful.

Though by the beginning of the Great Patriotic War the P-5, as well as the whole light reconnaissance - bomber class, had become obsolete, they continued to be employed (in night operations in particular) until 1944.

There were about 7,000 R-5s of different modifications produced. The aircraft displayed at the Air Force Museum was reconstructed in the city of Dushambe by a public design office, and was presented to the museum on February 2, 1993. The R-5 model is in excellent exposition form.

# The I-16

In 1933, N. Polikarpov's Design Office developed the I-16, an outstanding fighter in its time. It was not the first time that this office had designed a fighter aircraft of a monoplane design; a decade before N. Polokarpov and his colleagues had constructed the I-1 (Il-4000) fighter, which was even mass produced, but for a number of reasons it was not widely used (unsatisfactory spin quality, for instance).

Up to that time most of the fighters in our country and abroad (the Polikarpov's I-5 and I-15, in particular) had been of either a biplane or sesquiplane design.

The I-16, a cantilever monoplane with a roll-type fuselage and retractable landing gear, had a tightly cowled star-like engine: M-22 - in the first version, then the M-25 for a long period of time, and the M-62 and M-63 (1,1000 HP thrust) in the last versions.

The unusually compact dimensions of the aircraft attracted attention: span - 9 meters; length - 5.9 meters; wing area - 14.54 square meters; wing area load - 98-133 kg/sq.m. (according to weight in different modifications - from 1,400-1,940 kg).

The first flight of the experimental I-16, piloted by V. Chkalov, took place on December 30, 1933.

For a number of years after its appearance the I-16 had possessed the highest speed in the world. As a result of the continuous modifications, its max. speed increased from 440 kph in 1934 to 470 kph in 1940.

Its armament consisted of two 7.62 guns (in some modifications- four). There were also versions with two 20 mm guns.

In 1935 the I-16 was demonstrated at the Milan World exhibition and received world recognition. Practically each of its modifications possessed some new features which noticeably improved the I-16's flying characteristics.

The I-16s were produced for seven years, which is considered to be a rather long term for a fighter. In all, 6,555 planes were manufactured.

Since its first operations in Spain in late 1936, the I-16 had proved to be the world's most powerful fighter. It refuted the theory prevailing at that time that high-speed fighters were outclassed by low-speed machines. In fact, all depended on the type of maneuver.

It should be mentioned that the experience of air fights in Spain led to the disappearance of maneuverable biplane fighters.

The success of the I-16 operations in the skies of Spain made the opposing side modify its fighters rather urgently. The Messerschmitt 109 was replaced by the 109E, which was fitted with a more powerful engine (DB-601A at 1,025 HP, instead of 615-680 HP JuMO-210D) and developed speeds of 100 km/h and higher.

All that influenced the Soviet aviation to increase their fighter speeds. Yet, it was the I-16 that had greatly influenced the characteristics of fighters all over the world.

The I-16 was also employed in air combats in China, Mongolia (1937-39), and even during the Great Patriotic War up to 1944. Becoming obsolete by that time, the aircraft still contributed greatly to Soviet aviation achievements as it had been mastered by the flying personnel rather well.

It is noteworthy that the first "Guards" Air Force unit flew the I-16s.

The I-16 mock-up has been exhibited at the museum since February 1983. It was manufactured to order at the Kazan aircraft factory for the Air Force Museum.

# 3

## Aircraft of World War II and Post-War Propeller-Driven Aircraft

## The MiG-3 Fighter (Full-Scale Mock-up)

The experience of the combat operations in Spain and other local conflicts in the late 1930s provided for the intensive development of new aircraft. Several high-speed fighters were designed, quite new for those days, in several design offices. Some of them were chosen for mass production, among them the I-26 (Yak-1 in the future) in A. Yakovlev's Office; I-301 (the future LaGG-3), designed by S. Lavochkin, V. Gorbunov, and meters. Gudkov, and I-200 (the MiG-3) created by A. Mikoyan and meters. Gurevich.

The MiG-1 and its small-size modification MiG-3, which was included in the serial production, are monoplanes of composite construction (wood and metal), low wings (span - 10.3m; wing area - 17.44 square meters). The AM-35A engine installed in the aircraft developed a thrust of 1,350 HP at ground level and 1,200 HP at an altitude of over 7 km, which fully satisfied the designers in their intentions to create a high-altitude fighter, because before the war it was believed that all future air combats would take place mostly in the stratosphere.

It was decided to produce the plane at a definite aircraft factory (Number 1, taking into account the correlation of the production capacities of the wood processing and metal processing sectors of the factory, as well as the availability of the advanced technology. All those factors provided for a rapid mass production of the aircraft.

The first I-200 flight was performed by test-pilot A. Ekatov on April 5, 1940. The speed developed in the course of the tests was 651 km/h at an altitude of 7km and a ceiling of 12 km. The aircraft surpassed all world fighters of its time in speed, but at low and medium levels where the air combats were actually conducted its speed was much lower. Its armament - one 12.7 mm machine-gun and two 7.62 ShKAS machine-guns, proved not to be powerful enough in comparison to the wartime required cannon-equipped fighters.

Nevertheless, in November 1941 the MiG-3 serial production was stopped. The main reason for that was that the aircraft factories producing the AM-35A engines went over to producing similar AM-38 engines for Il-2 attack aircraft, which turned out to be the most urgent combat machine for the Soviet Air Force at that time. Besides, Il-2 had no analogs, though there existed other type fighters, such as the Jak-1 and LaGG-3. There appeared about 3,500 MiG-3 aircraft, a considerable part of them being employed in the Air Defense fighter aviation.

The museum displays a MiG-3 copy, designed by A. Mikoyan and meters. Gurevich. It was made at the "Zenith" machine-building plant and delivered to the Air Force Museum on 25th October, 1967.

# The Yak-9U Fighter

The Yak-9U was one of the last rotary-winged fighters designed by Yakovlev's Office.

During the war years the aircraft of this design office, beginning with the Yak-1 which first took to the air on January 13, 1940, piloted by test-pilot Yu. Piontkovsky, made up over half the fleet of the Soviet Air Force fighter division, with more than 36,000 of them being produced. The most mass produced among them was the Yak-9. All of the Yak fighters of the war-time were equipped with V. Klimov's engines, from VK-105P to VK-107A.

The Air Force Museum exhibits a Yak-9U aircraft with the 1,650 HP VK-107A engine.

The Yak-9U is of composite construction, with metal beams, veneer wing covering; duralumin landing flaps, aileron carcass, and tail unit, mass-balance of ailerons; fabric covering of the ailerons and the rudders; water radiator under the cabin; three blade VISh-61P airscrew; and a RSI-6 transmitting and receiving set. Its fuel tanks were malleable, protected with the neutral gas system: wing span was 9.74 meters; wing area - 17.15 square meters

The Yak-9U armament: one NS-37 gun and two UBS-12.7 machine-guns. The Yak-9U aircraft produced in 1944 with the VK-107A engine had a flight mass of 3,150 kg; speed - 680-700 km/h; range - 870 km; time to height of 5,000m -4.1 min, time to banked turn - 20 sec.

Nevertheless, such high flying performance wasn't practiced immediately due to engine imperfection. That's why only the Yak-9U with the 1,240 HP VK-105 PF-2 engine, providing speed over 620 kph was put into mass production.

The Yak-9U Air Force Museum exhibit was constructed at the aircraft factory in Novosibirsk on March 27, 1944 (No. 1257). The aircraft was flown in the Great Patriotic War combat operations. Later on it was reconstructed and delivered to the Air Force Museum on January 14, 1980.

# The La-7 Fighter

The La-7 fighter is a further modification of the LaGG-1 (I-301) family, designed by S. Lavochkin (co-designers V. Gorbunov and meters. Gudkov) in 1940. On March 30, 1940, test-pilot A. Nikashin performed the first LaGG-1 flight. The modified LaGG-1 was designated LaGG-3 and had been mass produced since July 1940, equipped with the VK-105P engine. The LaGG-3 powered by the ASh-82 and ASh-82FN engines was designated La-5. The La-5's worth as a fighter was proved in the Stalingrad Battle.

The modified La-5FN (La-7 prototype) was produced in December 1943, tested by N. Adamovich, and went to the production line in April 1944.

During the Great Patriotic War the aircraft industry produced over 22,000 La-type aircraft (LaGG-3-6, 500; La-5, 10,000; La-7, 5,753).

Together with the YAKs they formed the backbone of the Soviet Air Force fighter aviation in the war.

The La-7 aircraft were very popular among the flying personnel as these machines were characterized by an extremely successful combination of high-flying performance (the highest speed of our mass produced fighters), powerful armament, high survivability, and rather large fuel endurance.

They started delivering the La-7s to the front in the summer of 1944. Since the very first days they had been superior to the enemy Messerschmitt and Focke-Wulf aircraft.

At the final stage of the war, the La-7s were flown by the famous aces: N. Kozhedub, who downed 62 enemy aircraft; N. Skomorokhov, who personally shot down 46 enemy aircraft; and 8 other pilots within the group air combats (for example, K. Evstigneev, who downed 51 enemy planes, and others).

The La-7 is a single-seat fighter of composite construction and a low-wing monoplane. Span - 9.8 meters; wing area - 17.6 square meters; aircraft length - 8.6 meters.

The ASH-82FN 14-cylinder engine fixed in the aircraft develops max. thrust of 1,850 HP

The La-7 take-off weight is 3,265 kg. The max. speed at ground level - 600 km/h at the altitude of 6,000 m - 680 km/h.

Its armament: Two-three synchronized 20 mm guns. The ammunition: 260-390 shells. Two locks under the wing are designed for two bomb suspension, 100 kg each.

The exhibited La-7 was flown by N. Kozhedub at the front. He downed his last seventeen enemy aircraft in the Great Patriotic War in this plane.

The aircraft was delivered to the museum from the Central Aviation and Cosmonautics Club after meters. Frunze in July, 1960.

# The La-11 Fighter

In the post-war period S. Lavochkin Design Office continued improving piston-engine fighters. In 1946 the La-9 fighter was designed by them on the La-7 basis, and then there appeared the La-11 with the ASH-82 FN engine processing 1,850 HP take-off thrust.

The La-9 and the La-11 was all-metal, which considerably reduced the plane's empty weight. The wing had a more improved profile than that of a La-7, and no slats. The aircraft were equipped with the powerful armament of four 23 mm guns (300 shells) in the La-9 and three NS-23 guns in the La-11.

The fuel endurance and range increased. Besides, the advanced aerodynamics of the aircraft allowed it to develop a speed of 690 km/h.

The La-9 and La-11 were used as escort fighters. Due to the fuel tank capacity increase (up to 1,100) the La-11 range was 2,550 km.

After the factory and state tests the aircraft went into production, but those aircraft were only operated for a short period of time—they appeared as transitional aircraft because of the advent of jet fighters.

The La-11 aircraft exhibited at the museum was made at an aircraft factory in December 1948. Since 1948 it has been flown in a number of military units, with 296 flying hours in 436 flights. In February 1954 the aircraft was taken out of operation and transferred to the Moscow Aviation Technological Institute (MATI) where it was used as a teaching aid.

Later, the aircraft was dismantled, and on October 14, 1958, it was delivered to the Air Force Museum.

# The Il-2 Armored Attack Aircraft

In the period between the First and Second World Wars a number of states made attempts to create an attack aircraft capable of operating against selected ground targets from low altitudes while being exposed to active enemy fire. However, all the attempts failed as the method of "hanging" armor on the ready-made aircraft structure produced an overweighted aircraft possessing poor characteristics.

On the eve of World War II the problem of creating an attack aircraft became vital for the Soviet Air Force, taking into account the fact that the potential enemy had powerful armored units.

In 1938 S. Ilyushin submitted the idea of designing an aircraft with armor being an essential element of its structure.

Such an aircraft was designed and constructed in two experimental versions designated as UKB-55 (factory) and BSh-2 (Air Force).

Test-pilot V. Kokkinaki performed the first flights of both versions on October 2 and December 30, 1939, respectively.

The aircraft was a two-seat cantilever monoplane of composite structure with semi-retractable landing gear, powerful armament, and a streamlined armor fuselage accommodating the most essential parts of the plane: engine, crew, cabin, petrol, and oil tanks.

To meet the Air Force requirements the attack aircraft was modified into a single-seater (as its combat employment foresaw fighter escort). However, during the war there appeared an urgent need for its two-seater version as a result of severe combat experience: the unprotected aft hemisphere made the attack aircraft rather vulnerable to the Me-109 attacks, so the Il-2 single-seaters sustained heavy losses.

In early 1942 at a conference on the front, test-pilots of attack aircraft expressed the definite opinion that a gunner be introduced into the crew. The proposal was accepted and later on only two-seater versions were produced with a number of modifications: a more powerful AM-38F engine, stronger armament, a 15 degree sweepback of the leading edge to improve longitudinal stability, and replacement of the wooden outer wings to metal ones.

In all, during the war years there were produced over 36,000 Il-2 attack aircraft, and about 5,000 Il-10, its modification.

The Il-2 proved to be extremely effective as a combat machine. None of the fighting states' Air Force possessed an analogous attack aircraft.

The Il-2's main performance was the following in large-scale mass production modifications of the aircraft, such as Il-2M and Il-

1 (type 3): AM-38F engine, 1.750 HP thrust; span - 14.6 meters; wing area - 38.5 sq. meters; aircraft length - 11.6 meters; flight weight - 6,150-6.350 kg; max speed at ground level - 390-400 km/h, at an altitude of 1-1, 5 km/h; armament: two ShVAK 20 mm guns, later - VYa (23 mm) and, finally, NS-OKB-16 (13mm); two fixed ShKAS machine-guns and one UBT (12.7 mm); 4-8 rocket projectiles (RS-132 or KS-82, respectively) and 400-600 kg of bombs.

The IL-2 exhibit at the Air Force Museum was produced in October 1942 (No. 301060). The aircraft was engaged in the combat operations of the 243rd attack aircraft air division.

On December 30, 1942, pilot meters. Federov, returning from a combat mission in the attack aircraft, was hit by enemy fighters and made a forced belly landing in the peat bog called Nevniy Mokh, Novgorod Region. The plane was discovered there in 1977 and was delivered to Moscow. It was later reconstructed by S. Ilyushin's Design Office along with another downed Il-2 extracted from the peat bog. The reconstruction of the AM-38 engine was performed by the "Salyut" factory.

After the shooting of the films "Top Important Mission" ("Mosfilm" Studio) and "You Ought to Live" ("Lenfilm" studio), the Il-2 aircraft was delivered to the Air Force Museum.

# Il-10M Armored Attack Aircraft

Taking into consideration the Il-2¹/₂'s high combat effectiveness, in 1944 the Ilyushin Design Office developed another aircraft—the Il-10 advanced plane, powered by a AM-42 (1,750 kg HP) engine. It was an all-metal aircraft with improved aerodynamics, the present technology allowing for a more streamlined configuration. Having preserved the survivability of the Il-2 and the same powerful armament, the Il-2 surpassed its predecessor by 150 kph (speed at the altitude of 3,800 m was about 550 kph).

The Il-10 was on the production line from Autumn of 1944 till 1947, and was successfully operated during the final stages of the Great Patriotic War.

In 1951 a new modification, designated as Il-10M, was developed. It differed from the Il-10 in wing area, slotted flaps, more powerful armament, and increased mass. It also had a new tail unit arrangement that ensured better stability and controllability.

The first flights and production tests of the Il-10 and Il-10M, like other Il-family aircraft produced until the early 1960s, were performed by test-pilot V. Kokkinaki.

The Il-10M is an all-metal aircraft powered by a AM-42 engine. The armor is a hybrid of steel (4-16 mm thick) and duralumin (4-6mm).

The armament included four NR-23 fixed twin-wing mounted guns with 650 rounds; B-20 20 mm gun (150 rounds in electrified turret of gunner's compartment). Bombload - 400-600 kg of 1.5 to 100kg bombs in center-fuselage bomb bays; 2x250 kg bombs under wing and eight RS-82 or four RS-132 projectiles. Wing span - 14.0 meters; wing area - 33.0 square meters; aircraft length - 11.95 meters. Take-off weight, normal; - 7,100 kg maximum 0 7, 320 kg; max speed at the altitude of 2,650 m - 512 kph, at ground level - 476 kph; service ceiling - 7,000 meters; flight range - 1.070 km; take-off run - 440 meters; landing run - 500m.

The Il-10M was in serial production in 1953-54 and was operated till the 1960s. During the last years of its operation gunners for the Il-28 front-line bombers were trained in it.

The Il-10M was the last military piston-engine airplane series produced in the USSR.

The Il-10M exhibited was delivered to the AF Museum on June 17, 1959.

# Pe-2 Dive Bomber

V. Petlyakov, aircraft designer, was the closest collaborator of A. Tupolev, heading the wing development program of Tupolev's Design Office from ANT-1 to ANT-20. He launched the series production of ANT-6 (TB-3 heavy bomber). He also headed the work on the long-range high-altitude high-speed ANT-42 heavy bomber (first designated as TB-7 and then as Pe-8).

Under difficult circumstances, in prison, Petkyakov continued to work, and in 1938 he was assigned the task of designing a high-altitude high-speed twin-engine fighter-interceptor with a pressurized cabin, and such an aircraft (index 100) was made. Its maiden flight, performed by test-pilot P. Stefanovsky, took place on May 7, 1938.

The new aircraft tests were in full swing, but suddenly a government directive was received: in a short time the Design Office had to convert the fighter into a dive bomber. The task was rather complicated, but the office coped with it. The crew cabin, accommodating a pilot, navigator, and a gunner/radio operator, was not pressurized, and turbocompressors were replaced; two fixed machine-guns at the pilot's position, one turret machine-gun at the navigator's, and one turret machine-gun and a transportable one at the gunner's position were added. The machine-guns' caliber during production gradually increased from 7.62 to 12.7 mm.

The changes aimed at design simplification created the need to build a prototype: the dive bomber (designated Pe-2) was directly put into mass production in the late 1940s.

The Pe-2 was a low-wing vertical twin-finned monoplane. It was of an all-metal structure with a fabric covering of controls and ailerons. The aircraft was powered by two V. Klimov M-105RA engines, 1,100 HP each. The crew consisted of 3 persons.

Max speed was 540 kph; service ceiling - 8,800 meters; flight range - 1.300 km; wing span - 17 meters; wing area - 40.5 square meters; length overall - 12.66 meters; aircraft mass - 7,500-7,900 kg; bombload - 600-1,000 kg.

The Pe-2 trimmers, controlled airbrake, and engine cooling system flaps were controlled electronically.

The Pe-2 became the backbone of the Soviet front-bomber aviation of the Soviet Air Force during World War II. In all, 11,427 Pe-2s were built.

The Pe-2 had several versions, including the Pe-3 heavy two-seat fighter and Pe-2I powered by VK-107A engines (1,650 HP each), which developed a speed exceeding 650 kph. However, the Pe-3 was in a small series, and only several of the Pe-2I aircraft were built. There was also the Pe-2 prototype with Ash-82FN air-cooled 1,540 HP engines, but it did not see the production line.

During production the aircraft underwent some changes: VK-105R 1,100 HP engines were replaced by VK-105RF 1,260 HP engines, and a larger caliber armament was installed. In 1943, V. Myasishchev, who headed the Design Office after V. Petlyakov's death, redesigned the wing leading-edge into a less pointed configuration. These changes had to be introduced because initially the choice of the so-called high-speed V-profile was not a success. Though giving some gain in max speed, it also led to a landing speed increase, so take-off and landing characteristics were worse. That was an undesirable factor for operating the plane from limited size airfields. Besides, the V-V profile was the main reason for Pe-2 handling limitations, and the majority of pilots noticed it.

The Pe-2 exhibited was assembled from the remains of three planes, and was delivered to the Museum on July 14, 1959.

# Tu-2 Front-Line Bomber

The "103" aircraft (the future Tu-2) was designed under difficult conditions in a "closed" Design Office mainly manned by repressed specialists.

Designing began in March, 1940, and on January 29, 1941, the test-pilot meters. Myukhtikov performed the first flight. Powered by two AM-37 1,400 HP engines, the "103" aircraft displayed extremely high performance—the max speed of 635 km/h at an altitude of 8,000 meters was practically unattainable for all bombers of the time. It was easy to handle, and it possessed a bomb load of 1,000 kg to 3,000 kg (in overloaded version).

Simultaneously, with the beginning of series Tu-2 production, the AM-37 engine production was aborted because the production capacities of the plants manufacturing Mikulin engines were reoriented to production of AM-38 engines for the Il-2—the most required combat aircraft of that time.

The "103" aircraft was powered by ASh-82 1,700 HP engines (in the future replaced by ASh-82FN 1,850 HP ones). Simultaneously, one more crew member was added—the second gunner. Max speed became 100 km/h slower because of less altitude tolerance and a large engine "forehead." However, it was not of particular importance, as the speed normally lowered at high altitudes, but at the 1-4 km level typical for front-line bombers the speed loss was low. Besides, ASh-82 air-cooled engines provided higher bomber survivability than the AM-37 liquid-cooled ones.

The Tu-2 was an all-metal mid-wing monoplane with a twin-finned vertical tail.

Initially it was declared to be a front-line dive bomber, but in series production airbrake flaps were not installed as it was effectively used for level bombing normally.

Basic data of the Tu-2: wing span - 18.86 meters; wing area - 48.8 sq. meters; take-off weight normal - 10,860 kg, overloaded - 11,360 kg; max ground speed with ASh-82FN engines - 480 kph, at 5,400 m - 548 kph; service ceiling - 9,500 meters; flight range - 2,100 km; crew - 4 persons.

Armament: two 200 mm fixed guns with 150 rounds each to protect the forward hemisphere, and three 12.7 mm rotary machine-guns with 250 rounds each to protect the aft hemisphere. Bombload - 1,000-3,000 kg. Possessing such data, the Tu-2 was effectively used while sustaining low losses. It also proved to be easy to control.

The aircraft was the best front-line bomber of the World War II period.

Having many modifications, the Tu-2 series production was interrupted because the country was in dire need of fighters. However, in late 1943 the Tu-2 production started again.

As a result, only about 800 Tu-2s were produced and engaged in hostilities. The Tu-2 production went on after World War II - by the late 1940s 2,527 aircraft had been produced.

The Tu-2 exhibited in the AF Museum was delivered from an operational unit on November 21, 1958, in a very poor state.

Restoration took place at the production base of Tupolev's Design Office. In 1975 restoration work came to an end, and the aircraft was ready for display.

The exhibit is the only Tu-2 in Russia.

# Tu-4 Strategic Bomber

After World War II there was a need to develop a long-range strategic bomber which would meet the requirements of that time. The Tupolev Design Office designed the ANT-64 aircraft. The decision was made, however, not to construct that aircraft, but, instead, to make a replica of the American B-29 Superfortress, a strategic bomber possessing lower performance than that of the ANT-64, but accumulating the latest achievements of world aircraft technology, aircraft equipment (radars, in particular).

The main differences were as follows:

- the ASh-73 Soviet-made engines with TK-19 turbocompressors (A. Shvetsov Design Office) were used. It would have taken much time to copy the American Wright-Cyclone-3350-57 engines installed on B-29s, and besides, there was no advantage in power plant performance;
- the armament was replaced by Soviet-made ones—more powerful and advanced.
- the marking of dials on all visual instruments was made in metric units.

Solving the assigned task required a great number of materials, instruments, devices, systems, etc., and many novelties were introduced in metallurgy, chemistry, electrical, and radio equipment.

The development of the Tu-4 was very unusual. When the decision to build the aircraft was made in 1945, they began to construct not 2-3 prototypes, but 20 aircraft of small series. Those planes were production tested one after another with short intervals.

In May 1947, the first Tu-4 took to the air, operated by test-pilot N. Rybko's crew. They performed several development flights at the Kazan factory airfield. The main program tests were conducted at the base of a Flight Research Institute.

The second aircraft was flight-tested by meters. Gallai. The third crew commander was A. Vasilchenko. The tests lasted for about 2 years.

The Tu-4 is a mid-wing monoplane with three pressurized cabins. The machine is powered by four ASh-73TK-19 engines, 2,400 HP each. Later, a number of the Tu-4 were fitted with an in-flight refueling system.

The crew of the Tu-4 is 11-12 persons. Flight range - 6,500 km, speed - 558 kph, mass - 63,320 kg, service ceiling - 11.200 km. Armament: 8,000 kg of bombs, two large caliber machine-guns later replaced by two 23 mm cannons.

The Tu-4 exhibited (No. 280503), powered by ASh-73TK engines, was produced in March 1952, and was operated till 1958.

The aircraft logged over 1,540 hours and performed 2,004 landings. The last flight was carried out by Altukhov on October 7, 1958. It landed in Monino and entered the aircraft repair shop. After repairs, it became one of the Museum's first exhibits.

# Il-12 Airliner

World War II was in full swing. Il machines (attack aircraft and bombers) flew combat missions, but S. Ilyushin began designing a new passenger liner, having displayed his gift of engineering foresight. Unlike some other designers, S. Ilyushin decided to develop an airliner of specific structure and not to modify a bomber.

There existed no state agencies' requirements to the Il-12 liner, so Ilyushin himself formulated them.

The first Il-12 prototype, powered by two Charomsky ACh-31 diesel engines, was built in 1945. However, those engines were not reliable enough, so they had to be replaced by two ASh-82 FNs, and later by ASh-82Ts.

The Il-12 was a low-wing all-metal monoplane with fabric covering on controls and ailerons.

The passenger compartment and the crew cabin were provided with a heating system, ventilation, and were sound proof.

The landing gear was of the hydraulically activated tricycle type. The plane had a duel cable control. The wings, tail, screws, and pilot's forward window were fitted with an anti-icing system.

The first flight of the aircraft powered by ASh-82FN engines took place on January 9, 1946, followed by production flight tests. After those the Il-12 went to the production line.

Ground speed of the Il-12 at rated power was 350-375 kph, at an altitude of 2,250 m - 384-375 kph, flight speed at 3,000 m - 300-320 kph, service range at the same altitude - 1,150 km, and max range without payload could reach 3,230 km. The aircraft mass was: normal - 16,800 kg, overloaded - 17, 250 kg.

Depending on the range, the liner had several compartment arrangements. Basic versions could house 27 seats, and for short-range flights there was a 32 seat variant.

The Il-12 was series produced (1946-49) in different variations: passenger, transport, and a troop-carrier version that accommodated 37 paratroopers.

The Il-12 was widely used in Arctic and Antarctic expeditions.

The exhibited Il-12 (No. 30218) was constructed in February, 1948. In 1948-1959 it was flown by operational units. On February 17, 1959, the aircraft piloted by Saprikin landed in Monino and became part of the AF Museum exhibit.

# Il-14 Airliner

Work on the Il-14 started in late 1946. It was the further development of the Il-12. The aerodynamic characteristics and the arrangements were very similar to those of the Il-12. However, there were some novelties and improvements. The wing area was reduced and the thickness/chord ratio became larger, allowing for the placement of fuel tanks far from the passenger compartment. The vertical tail was enlarged to improve flight stability. Aerodynamic drag was reduced due to all three retractable landing gear unit bays being closed.

The aircraft was powered by ASh-82T engines with 1,900 HP take-off thrust.

All the required flight control and navigation equipment was installed, including ILS.

The first prototype took off in July, 1950. Production tests followed, at the final stage of which V. Kokkinaki took off with one running engine for the first time in the country's aviation history.

Production tests in 1952 were a success, and were followed by official state trials. Operational tests were conducted by Aeroflot and in a military transport aviation unit. As a result, the aircraft was found fit to be employed by Civil Aviation and the Air Force.

In Spring 1953, series production of the Il-14 began. It was constructed in various versions in the USSR and abroad. The basic version was the Il-14, housing 27-32 passengers. Flight speed at 3,000 m was 377 kph, and flight range at the same altitude with 26 passengers was 1,500 km.

The Il-14M passenger version had a lengthened fuselage that housed 39-40 seats.

The transport version—Il-14T—did not have any passenger seats and was constructed with an enlarged cargo door on the port side. Take-off weight was 17,500 kg.

The troop-carrier —Il-14D—had benches along the fuselage sides for paratroopers and necessary equipment.

A Photo-survey version could be flown in the Arctic.

Series production continued until 1957, and some of the versions were constructed till 1962. In 1987 Aeroflot and other countries' companies operated 270 Il-14s. Polar aviation operated these planes until the late 1980s.

The Il-14P exhibited (No. 146000610) was made in June, 1956. On July 5 it entered service with the 67th detachment of Civil Aviation. In 1965 it was flown to the Arkhangelsk detachment. During its operational life the aircraft logged 35,000 hours. It was delivered to the museum on August 15, 1974.

# An-2 Multi-Purpose Aircraft

The development of the plane began in 1946 when a biplane structure was considered to be obsolete. Yet O. Antonov, Chief Designer, had chosen the biplane with high mechanization, such as flaps on each wing, slats, and dropped ailerons, because the aircraft initially was to be used as an agricultural plane and was to be operated from small-size unprepared grounds.

The An-2 is an all-metal aircraft with fabric covered wings and tail.

The fuselage is of an all-metal semi-monocogue structure. The port side has a large cargo door (1.53x1.46m) with a smaller passenger door in it. Size of the passenger (cargo) compartment is 4.1x1.6x1.8m. The passenger compartment is sound-proof and can accommodate 12-14 light seats.

The first flight was made by test-pilot N. Volodin on August 31, 1947. The test flights of the agricultural version were conducted till July, 1948. For series production the An-2 powered by a ASh-62IR (1,000 HP) engine was selected. The aircraft was employed in agricultural, transport, and passenger versions.

Aircraft take-off weight normal - 5,250 kg, maximum - 5,500 kg; max speed at ground level - -253-255 kph, at 1,500 km - 253-268 kph, landing speed - 69-85 kph.

Operation of the An-2 started in August, 1948. Over 4,000,000 passengers were airlifted, and more than 90 percent of all air-chemical agricultural work was performed during its operational life.

Over 20 modifications were designed, and more than 10,000 aircraft were produced. The An-2 was the world's only aircraft series produced for over 40 years.

The aircraft (No. 70992) exhibited was produced on May 7, 1959. It logged 15,084 hours, made 26,707 landings, and had 10 repairs.

The aircraft operated on North-West airlines of the USSR: Pskov-Leningrad, Pskov-Tallinn. There were ambulance and forest patrolling missions as well.

On March 4, 1984, Egorov's crew landed the aircraft on Chkalovsky airfield, and again on April 5, 1984, in Monino, when it became an AF Museum exhibit.

# An-14 Multi-Purpose Aircraft

In 1958 the Design Office headed by O. Antonov developed the An-14, a liaison and multipurpose aircraft.

"Pchelka" (Little Bee) — is a strut braced high-wing monoplane with twin-fin vertical tail. High-life two-spar wings of cell type with automatically controlled extension flaps and drooped ailerons in combination with low wing loading provides steep trajectory of take-off and landing and steady gliding at low speed, as well as large attack angles. The aircraft could take off and land from very small grass airfields and pads. To take off in calm air a 100-110 m runway was enough, and by head-on wind - a 60-70 m runway.

Wing area - 39.72 square meters, max speed - 200 km/h, cruising speed - 170 km/h, max take-off weight - 3,750 kg, payload - 720 kg. The passenger version houses 7 persons (in compartment - 6 passengers, and one passenger and the pilot side by side).

In the agricultural version's fuselage there was a special tank for plant protecting chemicals or fertilizers.

The An-14 "Pchelka" (No. 500303) exhibited (powered by two Ivchenko AI-14F engines) was built in November, 1965, at Progress machine-building factory.

Production test flights were performed on June 1, 1966, with a flying time of 106 hours.

Then the flight tests were continued in a detachment of State Research and Development Institute of Civil Aviation.

On 18 March, 1967, routine flights began in Smolensk flying detachment.

Then the aircraft was operated in several military units.

On January 21, 1987, pilot Livinchev made the last flight of this aircraft and landed on the Monino aerodrome.

From the beginning of operational use the aircraft logged 700 hours and performed 677 landings.

# 4

## Subsonic and Transonic Jet Aircraft

### BI-1 Fighter

In the spring of 1941 engineers A. Bereznyak and A. Isayev (Bolkhovoitinov Experimental Design Office) began to develop the BI-1 fighter-interceptor powered by a L. Dushkin D-1A liquid-propellant engine with 1,100 kg thrust. The plane was to be fitted with two 20mm ShVAK guns.

That was the first USSR jet aircraft capable of performing independent flights from take-off to landing.

The BI-1 was an all-wood aircraft. Wing span - 6.48m; wing area - 7 square meters; length overall - 6.9 meters; take-off weight - 1,650 kg. Design max speed - over 800 kph, initial tests rate of climb - 82 mps.

Glider flights were followed by powered ones performed by V. Kudrin.

The first flight with a running liquid-propellant engine was performed by test-pilot G. Bakhchivandski on May 15, 1942. In all, 7 flights were carried out (one by K. Gruzdev). The flight time was over 6 minutes.

On March 27, 1943, trying to gain speed, the aircraft went into a dive and crashed.

Eight BIs were built, but none seen operational life, both because of the first prototype accident that was not cleared out and inadequate flight time for a combat fighter (low fuel endurance).

Liquid-propellant engines did not find application in aviation, but the accumulated experience of their development and operation was successfully used in spacecraft construction.

# MiG-9 Fighter

The MiG-9 (initially designated I-300) of A. Mikoyan and meters. Gurevich was designed together with the YAK-15 and was one of the first Soviet-made turbojets.

The MiG-9 was a single-seat, all-metal monoplane with a mid-mounted straight thin-section (9%) wing.

The first MiG-9 version was powered by two BMV-003 turbojet engines (800 kg HP each). These engines were replaced by RD-20 of the same structure. They were installed in the fuselage belly with jet propulsion nozzle exits towards the aft fuselage. The wing remained "clean." The bottom surface of the rear fuselage was pro-

tected against hot gases by a thin high-resistant steel shield. Landing gear was of the tricycle type.

The cabin and seat of the first aircraft were standard. The last MiG-9 modification cabin was pressurized and ventilated. Seats were of an ejection type.

Wing span - 10.0 meters; wing area - 18.2 square meters; length overall - 9.75 meters.

Armament: one N-37 gun with 40 rounds and two Ns-23 23-mm guns with 160 rounds each.

Take-off weight - about 5,000 kg; with external fuel tanks - about 5,500 kg.

The first flight of the MiG-9 was performed by test-pilot A. Grinchik on 24 April, 1946. The tests were a success and the speed was increased, but on June 11 during a demonstration flight the machine crashed.

Further tests of the MiG-9 were made by test-pilots meters. Gallai and G. Shiyanov. During one of the max speed flights performed by Gallai there was a partial structural failure which looked very much like that of Grinchik, but Gallai managed to land the aircraft. It was necessary to perfect the tail unit and specify the existing strength standards.

Further tests showed the increase of aircraft reliability. It became easy to control and accessible to a mid-skilled pilot. The decision was made to put the aircraft into series production. Thus the MiG-9 became a transitional aircraft: from propeller to jet thrust.

During tests speeds reached 410 km/h at an altitude of 5,000 meters.

Further speed increase required a swept wing. Flight range became 800 km without external tanks and 1,100 km with external tanks.

The MiG-9 (No. 114010) exhibited was built in 1947. On 15 November, 1947, Batyusky performed a production test flight.

Since 20 December, 1947, the MiG had seen operational use in several military units. The last flight was on 9 May, 1952 (pilot Solovyev).

From the beginning of operation the aircraft logged 133 hours. It was delivered to the museum on May 12, 1960.

# MiG-15 Fighter

As the possibilities of further speed increase of straight-wing aircraft were exhausted, in 1946 several aircraft Experimental Design Offices were assigned the task of designing a swept-wing fighter with a speed close to that of sound.

Such an aircraft (designated I-310), powered by the Rolls-Royce Mene licensed engine, was built in Mikoyan and Gurevich Experimental Design Office. The maiden flight was performed by test-pilot V. Yuganov on December 30, 1947.

After production and state tests, the MiG-15 aircraft was put into large-scale production.

The first aircraft modifications were powered by RD-45 (2,340 kg HP thrust) engines, which were built on the basis of Nene engines. The RD-45F (2,270 kg HP thrust) engine was used in production. After being modified by Klimov Design Office, the VK-1 (2,700 kg HP thrust) engine appeared, which powered not only MiG-15s, but many other types of aircraft.

The MiG-15 was a single-seat, all-metal, swept wing/tail transonic multipurpose monoplane. Wing leading-edge sweep angle - 35 degrees. Tail sweep vertical surface - 56 degrees, on horizontal - 40 degrees. Each outer wing plane has two stall fences. Wing span - 10.08 meters; wing aspect ratio - 4.85; wing area - 20.6 square meters; thickness/chord ratio - 10.3%

The fuselage is of a semimonocoque circular-section structure. In the rear part of the aircraft there is an engine. This rear position engine made it easy to replace. Landing gear is of the tricycle type. The cabin is pressurized and has an ejection seat.

The aircraft armament included: one NS-37 gun (with 40 rounds), two NS-23 guns (20 rounds each), two bombs, or two rocket packs. Guns were mounted in the forward fuselage under the cockpit.

During the production and state tests speeds reached 1,050 km/h. Service ceiling was 15,000 meters. It took the aircraft 2-3 min to climb to 5,000 meters. Take-off weight - 4,800 kg. Fuel system (with two 5001 external tanks) provided a flight range of 1,420 km.

After the tests the MiG-15 was put into mass production and entered service with fighter aviation of the USSR and many other states. The aircraft was available in a number of versions: front-line, fighter, escort fighter, interceptor, night-strike fighter, fighter-bomber, and photo-reconnaissance aircraft. In 1948 the MiG-15 was fitted with a VK-1 engine and was designated the MiG-15 bis.

It was of the same size and structure, but differed on some points. To control the ailerons, a reversible hydraulic actuator was used, and elevator balance was increased.

To eliminate wing dropping there appeared a "knife" stripe on the wing trailing edge. Moreover, the NS-23 guns were replaced by NR-23 high rate of fire guns.

The MiG-15 proved to be one of the world's best warplanes. It possessed high performance, powerful gun armament, high reliability, and was easy to handle. It really was a "work horse" and a "soldier aircraft", as it was called. For many years it was the backbone of Soviet fighter aviation.

The MiG-15 was the first Soviet jet fighter to be involved in the combat operations of the Korean War (1950-1955).

The MiG-15 exhibited (No. 2115368) powered by the VK-1 engine was produced in an attack fighter version with under-wing stores for bombs and rocket pods.

On December 13, 1957, test-pilot Chetverikov performed the first flight.

Since January 20, 1952, the aircraft has been operated by several units. The last flight was made my pilot Kurochkin on October 28, 1961.

In all, the plane flew 563 hours. On November 3, 1961, it became a Museum exhibit.

# MiG-15UTI Trainer

In 1949 the MiG-15 UTI two-seat trainer aircraft was created on the basis of the MiG-15. It had two separate cabins with a starboard-hinged canopy for the forward cockpit and a sliding canopy on the rear cockpit. Both cockpits had landing gear and flap controls. The aircraft also had ejection seats.

The MiG-15UTI (No. 22013) exhibited at the museum was built at a plant in 1954. On 26 July, 1954, production test-pilot Dotsenko performed the first flight. Since August 3, 1954, the air-craft has been operated at AF Research and Development Institute, as well as in several units.

Under the cosmonaut training program, pilot-cosmonauts Y. Gagarin, V. Tereshkova, A. Nikolaev, G. Beregovoy, and A. Leonov flew the aircraft.

On 5 May, 1973, pilot Vovk performed the last flight. From the beginning of its operational life flying time totaled 874 hours.

The aircraft was delivered to the AF Museum on 21 May, 1973.

# MiG-17 Fighter

The MiG-17 is a further development of the MiG-15bis with the same VK-1 engine and practically the same shape and structure. However, the wing is thinner, its area larger, and so is the sweep (45 degrees on leading edge and 42 degrees on wing tip). Wing span is 9.6 meters, wing area - 22.6 square meters. Take-off weight without external tanks -5,200 kg; with external tanks - 5,930 kg.

The MiG-17 has a longer fuselage. The cockpit is pressurized, air-conditioned, and partly armored from below. Front window is of armored glass. The ejection seat is of a new type, with the pilot's face protection curtain.

The last aircraft series elevators were fitted with irreversible hydraulic actuator.

Fuel was in the main rubber tank (1,250 l) aft of the cockpit, and provision was made for an additional metal tank (160 l) in the rear fuselage and two external tanks (400 l each). Armament: one N-37 gun with 40 rounds, and two NR-23s with 80 rounds each. The guns were of the MiG-15 arrangement. The aircraft could also carry two bombs on wing pylons.

By the end of 1949, three aircraft had been built. Number 1 aircraft was tested by I Ivashchenko. In February 1950 he reached a speed of 1,161 kph at an altitude of 5,000 meters. For the first time in the USSR a combat aircraft achieved sonic speed in a level flight. But, in March 1950, I. Ivashchenko perished in an aircraft accident. So, further tests were conducted on the Nos. 2 and 3 aircraft by G. Sedov and V. Kokkinaki. In mid- 1951 a large scale production of the MiG-17 was launched.

During the tests max speed was 1,114 kph, service ceiling, 15,600m. The height of 5,000 m was reached in 2 minutes. Flight range with external tanks was 1,735 km, without them - 1,300 km. Flight endurance with external tanks - 2 hours 53 minutes, take-off weight - 5,430 kg.

Until 1954 the MiG-17 was produced in the versions of a front-line fighter, fighter-interceptor, night interceptor, and photo-recon aircraft. In addition, there was a MiG-17F version powered by the VK-1F engine (3,380 kg HP with a 3-minute afterburning). Max speed increased up to 1,145 kph, service ceiling - 16,200 meters.

Combat performance of the MiG-17 proved to be high during the 1956 Middle East conflict when it countered the French "Mystere IV" fighter.

The exhibited aircraft (No. 1402001) was built in 1952 and was then delivered to the Research Agency.

The last flight was performed by S. Ilyushin on April 23, 1954. The aircraft made 160 flights with 86 hours flying time. At the end of its operational life it was delivered to the Air Force Academy as a training device.

On July 20, 1961, it became a museum exhibit.

# La-15 Fighter

From 1946 until 1948 the Lavochkin Design Office designed and produced several fighter prototypes, first straight-wing ones (La-150, La-153, La-156), and then—for the first time in the USSR—swept-wing fighters (La-160, La-168, La-174D).

In 1948, on the basis of the La-174D prototype, a front-line fighter (La-15) with the RD-500 engine (1,590 kg HP) was designed. It was a single-seat all-metal swept (37 degrees) high-wing monoplane. The horizontal tail was raised up. The tail unit and the wing were of the same sweep. The cabin was pressurized, and the landing gear was of the tricycle type. Armament: 3 x NS-23 guns (series planes had 2 ones).

In September 1948 the La-15, piloted by I. Fedorov, was successfully tested and recommended to production line.

During a test program speeds reached 1,026 kph at 3,000 meters, service ceiling - 14, 800 meters, flight range - 1,170 km.

The aircraft was serial produced and was in service with fighter units. However, it had never been in mass production because of manufacturing difficulties and low technological effectiveness of structure.

The La-15 exhibited was built in April, 1949. On April 23, test-pilot Alifanov performed the first flight under the production test program.

On May 20, 1949, the aircraft entered service with an operational unit, and in June it was delivered to the Air Force Research and Development Institute to be flight tested.

From the beginning of its operational use the flight time was 20 hours, and 37 landings were performed.

On May 23, 1950, the aircraft was delivered to the Zhukovsky AF Engineering Academy as a trainer, and in September 1958 became an exhibit of the AF museum.

# Su-25 Attack Aircraft

It is a single-seat attack aircraft intended for close air support of ground units. The development of the aircraft began in the Sukhoi Experimental Design Office in the early 1970s.

It is a monoplane with a high tapered wing with a low leading-edge sweep. It is powered by two R-195 turbojets with a total thrust of 9,000 kg HP.

Dimensions: wing span - 14.2 meters; wing area - 37.6 square meters; length overall - 15.2 meters.

When developing the Su-25, the Design Office paid close attention to meeting the specific requirements. The main requirements were, high combat effectiveness, simplicity in production and operation, low cost, and high combat survivability.

The latter was provided with an armored cockpit and armor over the most important units and systems; protected fuel tanks; pods of increased survivability and duel control systems; and fire protection of motor bays and compartments next to fuel tanks. The use of twin-engine arrangement played a substantial part in survivability increase.

To ensure high combat effectiveness the aircraft weapon system comprises a sighting complex with range finder/designator and weapons control system. Ten suspension points can carry from 100 to 500 kg bombs and pods with small-size cargo; laser-guided ASM, IR AAM; 57-370 mm rockets; pods with mobile 23-mm guns (with 200 rounds each). There is also an in-built 30 mm gun with 250 rounds.

The maiden flight of the Su-25 was performed by test-pilot V. Ilyushin in 1975. During the tests the following design data was recorded: max ground speed - 970 km/h. Max operational G-load - 6,5; Max combat load - 4,400 kg, normal combat load - 1,1460 kg, flight range with max combat load and two external tanks at ground level 750 km, at the altitude 0 1,250 km, service ceiling - 7,000 meters, take-off weight maximum - 17,600 kg, normal - 14,600 kg, take-off and landing runs - 600 meters.

Tests showed that the Su-25 was easy to handle, possessed high maneuverability, short turn radius, effectively used terrain for breakoff, etc.—i.e., it completely met the requirements.

While in service with the Air Force combat units, the Su-25 attack airplane proved to be a highly reliable aircraft.

The Su-25 (No. 25503101056) exhibited in the Museum was built in April of 1982.

On 24 May, 1982, the aircraft continued its test program. On June 21, 1989, pilot Komarnitsky performed the last flight of this aircraft. In all, the aircraft performed 622 flights and logged 445 hours.

On 15 September, 1990, the aircraft was delivered to the Museum from the Sukhoi's Experimental Design Office.

# Yak-17 Fighter

The Yak-17 is the further development of the Yak-15 fighter which had performed its first flight on 24 April, 1946, piloted by meters. Ivanov (the MiG-9 performed its first flight on the same day).

To reduce the period of development and make its production and operation easier it was decided to use the Yak-15 aircraft as a basis—the well-mastered and practice-approved Yak-3 structure having replaced the piston engine with an RD-10 jet engine (910 HP thrust). The Yak-15 had an arrangement (like the MiG-9) with exhaust gas flow under the rear fuselage, protected by a heat-proof shield. Such an approach proved its worth: the Yak-17 developed speeds of more than 800 km/h and was series produced.

Unlike the Yak-15, the Yak-17 had a nosewheel-type landing gear. The wing and fuselage required structural changes in order to retract the landing gear. These increased the take-off weight from 2,742 to 3,240 kg, and reduced max speeds to 751 km/h.

Service ceiling of the Yak-17 was 12,750 meters. Flight range - 717 km. Flight endurance - 1.6 hours. Armament: two 23 mm guns with 150 rounds. 430 aircraft were produced.

The aircraft also had a one-seat jet trainer version - Yak-17 UTI.

The exhibit was delivered to the museum on June 20, 1961.

# Yak-23 Fighter

The Yak-23 is the further development of the Yak-15 and -17 family. However, it is powered by a stronger RD-500 (Rolls Royce "Dernent" licensed) engine with 1,590 kg thrust.

The thin-section wing of the aircraft is straight, and the unpressurized cabin with 57 mm armored window has an ejection seat with 8 mm armored plate.

Tests were conducted by meters. Ivanov and were completed on September 12, 1947. Ground speed achieved was 932 kph, at 5,000 m - 913 kph, service ceiling - 15,000; fuel endurance with external tanks - 1,077 kg, take-off mass without external fuel - 2,900 kg. Flight range with external tanks - 1,300 km. Armament: 2 x 23 mm guns.

The Yak-23 was series produced. It was delivered to the AF Museum from the Yakolev Design Office on February 4, 1988.

# Yak-25 Interceptor

In the early 1950s A. Mikoyan, S. Lavochkin, and A. Yakovlev Experimental Design Offices were assigned the task of developing a night all-weather fighter-interceptor. The I-320, the La-200, and the Yak-25 were built, and underwent state tests. The first flight was performed on 19 June 1952 by test-pilot V. Smirnov. After the evaluation program the Yak-25 was adopted for service. It was a two-seat all-weather 35 degree sweep-wing interceptor. Powered by two Mikulin AM-5A wing-mounted engines (2,000 kg ho each), the Yak-25 reached a height of 15,250 meters at a maximum speed of 950 km/h (M = 0.9). Then, two RD-9 engines (2,600 kg HP each) were installed, and the Yak-25 increased its max speed to 1,090 km/h and service ceiling of to 16,500 meters. Flight range was 3,000 meters. Flight endurance - 2.5 hours. Take-off weight - 9,220 kg.

The crew consisted of a pilot and operator. The second crew member performed radar control.

Landing gear was of the bicycle type. Wing-mounted engines allowed the powerful "Sokol" (Falcon) radar to be installed in the "nose" fuselage.

The aircraft was armed with 2 x 37 mm guns.

The Yak-25 aircraft was serially produced and for a long time it was in service with AD air arm.

The exhibited Yak-25 (No. 0718) was produced in January, 1956. Its operational use began on 31 January, 1956. After production tests it was in service with a number of Air Force units. Pilot Dobrotullin performed its last flight on 10 July 1972, and the plane was delivered to the Museum.

The exhibit flew 1,592 hours, and 2,141 landings were made.

# Yak-25RV High-Altitude Reconnaissance Aircraft

In 1958 the Yak-25RV single-seat all-weather high-altitude recon aircraft powered by two R11V-300 turbojet engines (3,900 kg HP each) was designed and built. The first flight was performed on March 1, 1959, by test-pilot V. Smirnov.

Unlike the basic prototype, that aircraft possessed straight lengthened (about 10) wings. All that made it possible to reach a service ceiling of 21,050 meters. Max speed - 870 km/h. Flight range - 3,500 km. Flight endurance - up to 5 hours. Take-off weight - 9,000 kg.

International records were set up on the Yak-25RV: on 13 July 1959, Merited Test-Pilot V. Smirnov reached an altitude of 20,456 meters with a 1-ton load, and on 29 July he reached an altitude of 10,174 meters with a 2-ton load.

Two international women's records were set by test-pilot meters. Popovich. On 11 August 1965 she gained a speed of 735 km/h over 2,000 km closed circuit. On 18 September 1967 she set a flight distance record (2,497 km) over closed circuit.

The Yak-25RV (No. 25992004) exhibited in the museum was produced in November, 1965.

Test-pilot V. Smirnov performed the first test flight on 4 November 1965, and on 6 November 1965 he carried out the flight-acceptance test.

On 21 December 1965, the aircraft was delivered to the Flight Research agency, where test flights were performed with a total flying time of 186 hours. On 9 July 1969 the Yak-plane flew to another agency where flights proceeded. In all, the aircraft logged 233 hours. The last flight was performed by pilot Volk on 15 June 1973. The aircraft landed on Monino airfield and became an exhibit.

# Yak-36 VTOL Fighter

VTOL aircraft are designed to combine the capability of off-field operation ("from the point") with the availability of combat aircraft max speed. The design of the Yak-36 aircraft began in the 1960s. The power plant used provided both vertical take-off and level flight.

For this purpose two R-27-300 turbojet engines were installed in jet nozzles of which there were blades turning propulsive jets.

Aircraft control during vertical take-off and landing, as well as at transient regime, is provided by control jets. There are air jet nozzles located on the front edge of the nose strut, in the rear fuselage, and on the wingtips. Compressed air flows from engine compressors to the exhaust nozzles and ejects with high velocity. Thus, forces are created which provide aircraft trim of lateral and longi-

tudinal axes. Standard aerodynamic controls come into operation after the boost.

The Yak-36 is a high-wing monoplane. On 16 September 1963, test-pilot Yu Garnaev performed the maiden flight. In August 1964, V. Mukhin, a test-pilot of the Yakovlev Experimental Design Office, began flying the aircraft. These pilots carried out the majority of the test flights.

During the test program flight speeds reached 1,000 km/h.

The Yak-36 aircraft was displayed at Domodedovo Air Show on 9 July 1967. Demonstration flights were performed by test-pilot V. Mukhin.

Mukhin piloted this airplane for the last time on 10 July, 1967. In all, 270 flights were carried out, and the flying time total was 23 hours.

# Yak-38 Carrier-Based Aircraft

The Yak-38 is designed for destroying sea and ground targets. It is a mid-wing monoplane with a swept wing. To decrease the size, wingtips fold upward at 102 degrees. The aircraft has tricycle landing gear. The first flight was performed by test-pilot V. Mukhin on January 15, 1970.

The aircraft control at vertical and transition regimes is provided by control jets which are similar to those of the Yak-36, but, unlike the latter, there is no nose strut on the Yak-38, and the front nozzle is on the nose fuselage.

For emergency escape the K-36VM ejection seat is used. Ejection is performed manually with release of the canopy at vertical and transient regimes—through the window manually, or automatically at critical and roll parameters.

The power plant comprises one lift/cruise engine and two lift engines. The R27B-300 lift/cruise engine is in the center fuselage, and it has a vertical take-off thrust of 5,900 kg HP (max afterburning thrust - 6,600 kg HP).

The RD36-35FV lift-jets are installed vertically in the front fuselage behind the cockpit.

Each lift engine weighs about 250 kg and has 2,890 kg HP of thrust. Uninterrupted work time is 2 min. When using only the lift/cruise engine the aircraft can perform conventional take-off from and landing on a concrete runway.

The Yak-38 aircraft is equipped with a Kvadrat-D (Square-D) short-range guidance system.

Armament: 6x100 kg bombs; two R-60 and two H-23 missiles; two LPK-23 pods; rockets; and Gsh-23L gun in fighter version. Total combat load - 700 kg.

Max flight speed - 1,210 km/h, service ceiling - 12,300 meters, ground level flight range - 400-500 km, at the altitude of 10,000 m - 880 km, tactical radius at the altitude of 10,000, at the speed of 950 km/h - 195 km. Take-off weight - 10,300 kg, fuel endurance of two tanks - 2,900 kg. Crew - one pilot.

In 1972 the Yak-28 performed deck landing on the "Moskva" cruiser. At present the aircraft is widely used by "Kiev", "Baku", "Minsk", and "Novorossiisk" antisubmarine aircraft-carrier cruisers.

The Yak-38 exhibited was built in March 1975. Test-pilot Dexbuch performed the maiden flight on 31 March 1975. Since 1975 the aircraft has been operated in several units. In all, 70 flights were made, with a flying time of 20 hours. On 24 April, 1978 it was delivered to the Zhukovsky Air Force Engineering Academy training base, and in 1989 it became an Air Force Museum exhibit.

# Tu-16 Bomber

The Tu-16 (Tupolev Design Office designation—"Aircraft 88") is a mid-wing monoplane with 35 degree wing sweep. It is powered by two AM-3M engines, 8,750 kg HP each. Armament: seven 23 mm guns; bombload - 3-9 tons. Crew - 6 persons. Aircraft mass - 76,000 kg; max speed - 1,020 kph, landing speed - 230 kph, service ceiling - 13,000; flight range - 5,800-6,400 km; wing span - 32.9 meters; length overall - 34.8; wing area - 164.7 square meters.

The first Tu-16 prototype flight was performed by test-pilot N. Rybko on April 27, 1952.

The aircraft has been series produced in several versions for many years, not only in the USSR, but in the Chinese People's Republic as well.

NATO's code name is "Badger."

The Tu-16 exhibited (No. 18800302) powered by the RD-3M engine was produced in 1954. Production tests were made on December 18, 1954, by test-pilot A. Kazakov. In 1955 the aircraft flew to the AF Research and Development Institute where test flights continued until August 1960. On August 6, 1961, the aircraft was delivered to the Zhukovsky AF Engineering Academy training base, and later it became an exhibit of the AF Museum.

The aircraft had logged 562 hours 18 minutes and made 218 landings.

The AF Museum also exhibits the Tu-16K version, a missile carrier, which can carry two ASMs.

The aircraft's last flight was performed by Pilot Pavlenko on June 15, 1966. In all, the plane had logged 1,550 hours 10 minutes and performed 1,122 landings.

In May 1967 the aircraft was delivered to the training base of Zhukovsky Air Force Engineering Academy. Later it became an AF Museum exhibit.

# Tu-104 Airliner

The Tu-104 originated from the Tu-16 bomber which proved its worth in operation. Wings, power plant, tail unit, main landing gear, and substantial parts of aircraft systems and equipment remained practically unchanged. The volume of the pressurized fuselage was enlarged and raised, so the liner turned into a low-wing monoplane.

Naturally, having originated from a bomber, the liner possessed lower economic characteristics than a specially designed airliner could have had. However, the Tu-16 gained operational experience and this resulted in a higher reliability in the Tu-104. That was a

very important factor in developing the country's first multi-seat jet airliner, which gave practically a three-fold leap in the passenger transportation rate. Besides, in general that concept made it possible to solve the assigned task in a very short period of time, so the decision proved worthwhile.

The first Tu-104 prototype flight was performed by test-pilot Yu. Alasheyev on June 17, 1955, and routine flights began on September 15, 1956, when the liner flew from Moscow to Irkutsk. That flight paved the way to the world's jet passenger route.

The first Tu-104 accommodated 50 passengers, the Tu-104A - 70, and the Tu-104B - 100 people. The last series planes accommodated 115 passengers. The Tu-104 had set 26 world records in speed and load-carrying capability from 1957-1959. That number of jet airliner records was second to none.

In 1958 at the Brussels International Show Tu-104 gained a Gold Medal.

The Tu-104A exhibited (No. 8350705) was constructed in 1958. On December 28 pilot Korobko performed production tests.

On January 28, 1959, that aircraft entered the service with a special operations detachment of Civil Aviation, and from June 25,

1962, it was with the 200 detachment of Civil Aviation. From June 6, 1970, it was operated by the 207 detachment.

The aircraft had logged 9,851 hours 14 minutes and made 5,051 landings.

After a certain refurbishing the Tu-104A was employed for cosmonaut training in zero-G conditions of short duration (2,313 flights on weightlessness were carried out).

The aircraft was delivered to the AF Museum on January 16, 1979, from a military unit, having finished its operational life.

# Tu-95 Intercontinental Missile-Carrier

The aircraft was designed to meet the requirements of a strategic bomber with range exceeding 10,000 km. It took the designers only eight months to work out the design. Some six months after, the flight tests began.

On November 11, 1952, the Tu-95 took to the air piloted by A. Perelyet, who initiated the test program. However, on May 11, 1953, one of the two TV-2F engines powering the craft (NK-12 was not ready yet), caught fire in flight. The crew did its best to prevent an accident with no success. A. Perelyet ordered the crew members to escape. He and the mechanic continued to fight the fire, and unfortunately perished: the control rods seem to have been burnt out.

No. 2 aircraft was constructed and fitted with four turboprop NK-2 engines, 15,000 kg HP each. The maiden flight of the aircraft was made on February 16, 1955. The time between the accident and No. 2's maiden flight was not wasted, but used for an intensive program of full-scale tests of the NK-12 engine. The tests were carried out on board the Tu-4ZZ flying laboratory, where one of the starboard engines was replaced by NK-12. That unique job was performed by the crew headed by test-pilot meters. Nyukhtikov and leading engineer D. Kantor.

It was meters. Nyukhtikov who took the second Tu-95 aloft and completed the tests. The aircraft went to the production line, and is still in service with long-range aviation units—an example of unusual longevity.

The crew of the Tu-95 consists of eight men. Wing span is 54.0 meters; wing area - 310.05 sq. meters; length overall - 47 meters.

Speed of the machine is 910 kph; take-off weight - 182 tons; service ceiling - 12,000 meters; range (of further modifications) - up to 18,000 meters; fuel endurance - up to 94,000 kg.

In 1960 the aircraft was converted into a flying laboratory, having logged over 369 hours and performed 224 landings.

The Tu-95B display was delivered to the AF Museum on June 17, 1959.

## Tu-114 Airliner

The Tu-104 originated from the Tu-16 bomber, so the Tu-95 bomber served as the basis for the Tu-114 airliner, which inherited its wing, tail unit, powerplant (four NK-12 engines with 15,000 kg HP each), main landing gear, and a substantial part of its equipment. The new designed fuselage accommodated a 160-220 seat pressurized passenger compartment.

The first flight (November 15, 1957) and the basic tests were conducted by test-pilot A. Yakimov's crew. In December 1959 the aircraft was removed to the Research and Development Institute.

The Tu-114 basic data: speed max - 810 kph; cruising speed - 770 kph; range - up to 8,000 km; take-off mass - 171,000kg.

The aircraft was small-series produced, but was long used on domestic and international airlines, having displayed high reliability: no accidents through engineering equipment failure.

In 1961-62 the aircraft set 32 world records. It earned the "Grand Prix" at the Brussels International Show, and FAI awarded the Gold Medal to A. Tupolev.

The Tu-114 exhibited was manufactured on October, 1957.

From May 1963 till September 1965 the aircraft underwent some development improvements at Kuibyshev machine-building factory, after which its operational life reached 500 flying hours.

The aircraft was flown till 1972 and had logged 794 hours 46 minutes.

On March 16, 1972, pilot I. Sukhomlin landed the aircraft on the Monino airfield.

# Il-28 Front-Line Bomber

In the late 1940s some of the Design Offices were set the task of developing a front-line bomber, powered by two turbojets. By that time everyone realized the advantages of a swept wing at high transonic speeds. Nevertheless, S. Ilyushin made the decision to use a straight wing. The decision could have been regarded as a rather conservative one, but in fact it was well grounded: such a wing provided better performance at design speeds of Mach 0.8.

The Il-28 was a high-wing monoplane with a tapered straight along the leading edge wing and swept single-fin tail. The wing span was 24.45 meters; wing area - 60.8; length overall - 17.65 meters. The all-metal structure of the aircraft was of high technology.

The crew consisted of three men: pilot, navigator, and tail gunner. Both front and rear cabins were pressurized and armor plated. Landing gear was of tricycle type with a nose wheel.

The plane was powered by two VK-1A engines, 2,700 kg HP each, mounted underwing. The five tanks' capacity was 7,908 l.

The aircraft was armed with 2 NR-23 fixed guns (200 rounds) mounted in the lower nose fuselage, and two NR-23 movable guns (450 rounds) in the tail section.

Bombload normal to 1,000 kg, maximum - up to 3,000 kg in fuselage bomb bay.

The normal take-off weight - 18,400 kg; max - 23,200 kg; empty weight - 12,890 kg.

The aircraft's first flight was performed by test-pilot V. Kokkinaki on July 8, 1948. The test program showed the advantages of the aircraft in comparison with swept-wing aircraft of the same function developed by other design offices. It went to the production line in various versions: front-line bomber; recon aircraft with additional fuel tanks; torpedo-carrier with two torpedoes in a lengthened bomb bay; trainer; and some others.

In its front-line bomber version developed a max speed exceeding 900 kph at 4,000 meters; service ceiling was 12,500 meters; range - up to 2,400 km.

The simplicity of handling, structural strength, and reliability were highly appreciated by pilots.

The museum exhibit was manufactured in December 1953. It went through the production test program.

Since February 9, 1954, the aircraft had been operated by various units. After repairs (July, 1958) it continued its operational life (until January 1965).

In all, the aircraft logged 1,296 hours in 1,184 flights.

On January 15, 1965, pilot Slyshchenko landed the aircraft in Monino and the Museum acquired a new exhibit.

# Il-18 Liner

The four turboprop-engined Il-18s (not to be confused with the earlier constructed piston-engine plane) were the first Soviet-made liners, and not simply modified bombers. The Il-18 is an all-metal classic low-wing monoplane with double-slotted flaps. The capacity of pressurized wing fuel compartments is 23,000 lt.

The pressurized fuselage provides the necessary comfort for the passengers. The conditioner system guarantees the temperature of about 20 degrees C. The pressure inside the passenger compartment at the altitude of 5,200 m corresponds approximately to that of the ground.

The four AI-20 turboprop engines, designed by A. Ivanchenko, are attached to the wings two by two. The power of each engine is equivalent to 4,000 HP.

The aircraft is fitted with flight-control and navigation equipment, which ensures flights in poor weather conditions. Il-18 was the first home-made aircraft to be equipped with an automated flight control system, incorporating the control system and autopilot.

The first Il-18 flight was performed by V. Kokkinaki on July 4, 1957, followed by production tests. In the course of those tests a flight from Moscow-Irkutsk-Petropavlovsk-Kamchatka-Tiksi Bay-North Pole 6 Station-Tiksi Bay and back to Moscow was performed. In 27 hours the aircraft covered a distance of 18,000 km at an average speed of 650 kph. That flight demonstrated the high performance and maintenance qualities of the plane.

From May to August 1958 the aircraft successfully went through state trials and went to the production line.

The first aircraft were constructed in 89 passenger variant, their commercial cargo being 12,000 kg and the take-off weight about 58,000 kg. Those figures were soon improved.

The last Il-18 modification was on the production line from 1965 until 1969; its take-off mass was 64,000 kg; it could accommodate 122 passengers; and the cruising speed was 650 km/h; flight range with maximum commercial cargo (13,500 kg) - 3,700 km. It was a 5 man crew aircraft.

The Il-18 production of various versions continued until April 17, 1969. The total number of produced aircraft equaled 564 planes. The Il-18 was the most massively produced aircraft, having been the first Soviet-made airliner which had entered the world market.

The aircraft was demonstrated at the International Air Show in Brussels, flew to the majority of the world's countries, and to the Arctic and Anarctic Continents.

In 1959-69 the airliner set 22 official world records in speed, range, and weight capacity.

The Il-18 aircraft (No. 181002701) exhibited in the Museum was constructed at the "Znamya Truda" (The Banner of Labour) aircraft factory. It performed 13,718 flights and logged 34,998 hours.

On July 12, 1977, this plane carried out its last take-off from one of the Leningrad airfields, landed in Monino, and became an exhibit of the Air Force Museum.

# Il-62 Airliner

By the end of the 1950s the design offices of our country had begun work on the next generation of jet airliners, powered by by-pass turbojets. In 1960 the Ilyushin Office began to work on the Il-62 passenger aircraft. It was a low-sweep-wing monoplane (length 53.12 m). The wing area was 280 square meters and span - 43.2 meters. The leading edge was stepped and the wing span profile varied. These provided for longitudinal stability of the aircraft in all angles-of-attack, so flight safety in atmospheric turbulence zones was achieved.

The powerplant included four Kuznetsov NK-8 engines with a maximum thrust of 9,500 kg each. Later, the NK-8 thrust rose to 10,500 kg, and the designation was NK-8-4. A thrust reverse reduced the landing run. The Il-62M (modified in 1969) was equipped with more economical and powerful (11,000 kg) D-30KU by-pass turbojets designed by P. Soloviev. The engines were attached to the tail unit. Such an arrangement provided a "pure" wing with high aerodynamic features and effective mechanization. The engines were close to the longitudinal axis, preventing the aircraft from unbalance in case of engine failure. The passengers and the crew were not disturbed by noise. Fire safety was also higher.

The maiden flight of the Il-62 prototype was performed by V. Kokkinaki, a test-pilot, on January 3, 1963. In the test flights the following data were obtained: the maximum take-off weight with a commercial load of about 23,000 kg made from 157,500 kg to 165,000 kg the practical range at a cruising speed of 850 kph having originally been from 6,700 to 8,300 km; max range with 10,000 kg load - 9,200 km; max speed - 950 kph; 5-man crew. The aircraft could accommodate 168-186 passengers.

The first regular Il-62 passenger flights started from Moscow to Alma-Ata on September 8, 1967. A little bit later a technical flight from Moscow-Petropavlovsk-Kamchatka was made, followed by regular flights to Kamchatka and other Far East towns. In 1971 Il-62 appeared on world air lines.

The Il-62 was demonstrated twice at Paris International Shows. It was recognized all over the world, and 55 planes of the type were operated by other countries. That was the last aircraft designed and modified by S. Ilyushin. After S. Ilyushin's death the Design Office was headed by G. Novozhilov.

The Il-62 (No. 70205) exhibited in the Museum was constructed in September, 1967. On September 28 it was first flown by a production test-pilot.

In December 1967 it entered service with the Domodedovo air detachment and its regular cruises began.

In 4,288 flights the aircraft logged 14, 891 hours. The last flight was flown in May 1981, and on May 16, 1983, it landed on the Museum runway.

# 3M Strategic Bomber

The 3M strategic intercontinental jet bomber was a further development of Miasishchev M-4 bomber.

The elaboration of the M-4 was a real breakthrough in the development of the aircraft for such functions. From the beginning it was considered impossible to bring in line high transonic maximum speed (unattainable without turbojets), intercontinental range, and nuclear-bomb load-carrying capability. But the difficulties were overcome due to many technological novelties.

The 35 degree sweep wing was lengthened and highly flexible. Four A. Mikulin AM-3M engines, 8,750 kg thrust each, were attached to the wing root two by two close to the fuselage. The bicycle-type landing gear had a "ramping" nose strut (ramping automatically after the required take-off run speed was gained). Aircraft control was power-operated and irreversible. Crew ejection was performed downward. The bomb bay could accommodate any bomb in service.

Small arms defense weapons included eight 23 mm twin remotely controlled rotary cannons.

The avionics made it possible to operate the aircraft both day and night under poor weather conditions.

In January 1953 the 103M took to the air. It was flown by F. Opadchiy, a test-pilot. After the test flights the aircraft went to the production line, but its performance, stability, control, (especially during the take-off) and flight safety required further development.

So, V. Miasishchev's Office designed the 3M plane having modified the profile, changed the nose, and horizontal tail. The wing span was expanded. The Design Office also used new materials. The 3M take-off weight was 202 tons. The aircraft was powered by four VD-7 engines (designer V. Dobrynin), with 11,000 kg thrust each. Hydraulic unit of "ramping" the nose strut was replaced by a pneumatic one. The aim was to provide a more smooth angle-of-attack increase during lift-off. The wing span was 52 meters, the length - 49 meters. On March 27, 1956, the 3M made its maiden flight, performed by the crew of test-pilot meters. Gallai.

The speed achieved during the flight was 900 kph; range - 9,500 km; ceiling - 15,600 meters; the bombload - up to 18,000 kg.

19 world records were set by the M-4, and 3M, and on September 16 N. Goryainov's crew took off with the weight of 10 tons to the altitude of 15,317 meters.

In September 1959, A. Lipko's crew performed the Moscow-Orsha-Moscow flight having covered the distance of 1,000 km with the weight a of 27 tons at the speed of 1,028.6 kph. On October 20, 1959, B. Stepanov's crew gained the altitude of 13,121 m with 55-ton weight.

The modified 3M (designated TM) airlifted the "Buran" space vehicle, and later the bulky Energiya missile fuel tank from Moscow to Baikonur. That mission required a number of modifications: the single-fin tail was replaced by a twin-finned one.

The 3M exhibited in the AF Museum (No. 0301804) was constructed in 1960. From May 10, 1961, it was operated in different units. In all, the aircraft had logged 4,904 hours 52 minutes and performed 2,384 landings. In July 1986 the aircraft landed in Monino and became an exhibit.

# Experimental M-17 High-Altitude Aircraft

In 1982 the Experimental factory, named after Miasishchev and under the leadership of the Chief Designer V. Novikov, developed the M-17 high-altitude subsonic prototype, the only one in the Soviet Union. Only several aircraft of its kind existed in the world. Now atmospheric research of long duration can be performed at altitudes of up to 22 km. High speed and long range allow for missions that require observation of vast areas. The aircraft has special equipment weighing 1,000 kg for examining the ozone layer of the Earth's atmosphere. The first M-17 took off on May 26, 1982, piloted by E. Cheltsov.

The aircraft is an all-metal cantilever high-wing monoplane. It has two fins and a high horizontal tail.

The thrust of the RD-36A engine is 7,000 kg.

The aircraft length is 21.085 meters; height - 4,874 meters; the wing span - 40.3 meters; the empty weight - 14,318 kg.

The exhibited M-17 aircraft performed 187 sorties. It was flown to the AF Museum by test-pilot Generalov from Zhukovsky township on January 25, 1990.

# Yak-40 Airliner

In the early 1960s the Yakovlev Design Office developed a relatively small size airliner capable of operating from the grass strips of local air lines.

On October 21, 1966, the aircraft piloted by A. Kolessov took off. That was Yak-40, a straight low-wing aircraft powered by three by-pass AI-25 engines with 1,500 kg HP each. All engines are in the rear part; two on the sides of the fuselage and the third one inside the fuselage.

The compartment could house 24 passengers at first, and was expanded to 32 seats and is equipped with an in-built stairway.

Dimensions: wing span - 25 meters; wing area - 70 square meters; length - 20.36.

In 1967 Yak-40 went through state tests having gained the cruising speed of 550 kph. The takc-off mass was 16,100 kg; max commercial load - 2,720 kg; range with max commercial load - 1,200 km, with 2,100 kg - up to 1,800 km. The crew - 2 pilots.

The Yak-40 was widely used on national airlines and by some foreign companies.

The Yak-40 (No. 9110117), exhibited in the Museum, was constructed on March 1971. On March 17 test-pilot Rabota performed a production test flight. On April 15 the plane was delivered to an organization where various tests and demonstration flights were carried out.

In all, the aircraft logged 1,428 hours. In its last flight the aircraft piloted by Lyagushkin landed in Monino. It has been in the inventory of the Museum since January 26, 1981.

# Yak-42 Liner

The Yak-42 swept low-wing aircraft was designed and constructed in 1975.

The powerplant includes three D-36 bypass turbojets (designer P. Soloviev), 6,500 kg HP each, in a rear fuselage arrangement: two on the sides and one inside the fuselage. Such an arrangement provided comfortable conditions in the passenger compartment.

There are 120 seats in the compartment. To speed up passenger exit and entry the aircraft has a built-in stairway. On March 6, 1975, test flights were started, and were performed by A. Kolesov. The aircraft went to the production line in that same year.

Dimensions: length - 36.38 meters; wing span - 34.88 meters; wing area - 150 square meters; wing sweep - 25 degrees.

During the test flights the aircraft gained the cruising speed of 810 kph. Take-off mass - 53,500 kg. Max commercial load - 14,500 kg; range with max commercial load - 1,000 km, with a load of 10,500 kg - 1,850 km. A crew of two.

The exhibited aircraft (No. 01-004) was constructed in May, 1978. On May 17 it was test flown by test-pilot Shevyakov and was recognized as operable.

On June 1, 1978, the aircraft was delivered to an organization to be flight tested.

The plane logged 178 hours. It landed on the Monino airfield on February 11, 1981, piloted by Lyagushkin.

# An-8 Military Transport Aircraft

The An-8 is a straight high-wing aircraft with broad fuselages and two AI-20D turboprop engines with 5,180 kg HP each (designer A. Ivchenko). The nacelles for main gear retraction are closely attached on the sides of the fuselage. The construction of the An-8 solved a complicated problem - air-landing of bulky hardware. The aircraft has a large cargo compartment (11x2, 4x2, 5) with a freight hatch in the rear fuselage. The cargo compartment is fitted with loading/unloading equipment.

The An-8 was a high-performance plane, the undercarriage-high-passable. It could operate from grass runways.

Dimensions: length - 30.74 meters; wing span - 37 meters; height - 10.045 meters; wing area - 117.2 square meters; take-off run - 700 meters; landing run - 450 meters; empty mass - 24,600 kg; take-off mass - 38,100 kg; max load - 11,000 kg; mass speed up to 580 km/h; cruising speed - 450 km/h; landing speed - 175 km/h;

service ceiling - 10,500 meters; range with full load - 400 km, with the 5-ton load - up to 2,800 meters.

The exhibited An-8 (No. 9340504) was constructed at the aircraft factory on September 1, 1959.

The first flight was performed (by Ya. Vernikov crew) on February 11, 1955.

In all, the aircraft had logged 3350 flying hours, having performed 3450 landings. It was delivered to the Museum from a military unit on May 20, 1976.

# An-10A Airliner

The An-10A is a modification of An-10. Its fuselage is 1.1 m longer. The passenger compartment was re-arranged and more seats were accommodated, which increased the economic efficiency of the plane.

The An-10A is an all-metal high-wing with broad fuselage, low ten-wheel undercarriage. It is powered by four AI-20K turbo-props, 4,000 kg HP each (Chief Designer A. Ivchenko).

The first flight was performed by the test-pilot Ya. Vernikov and crew.

Performance cruising speed at the altitude of 8,000 m - 630-650 kph; service ceiling - 10,200 meters; landing speed - 190-200 kph; take-off run - 700-800 meters; landing run - 600-650 meters. Empty mass 32,000 kg; mas mass - 54,000 kg; passenger number - over 100.

The exhibited aircraft was constructed at an aircraft factory in 1960 (No. 0402406).

On January 18, 1961, the aircraft entered the service with the 75th Detachment of Civil Aviation.

It had logged 16,360 flying hours (after last repairing - 2,271 hours).

On January 14, 1976, the plane was delivered to the Museum from the Siktivkar air detachment.

# An-12 Military Transport Aircraft Office

In 1957 the Office headed by Chief Designer O. Antonov developed An-12 transport aircraft designed for airlifting bulky cargo. It could also perform parachute drops and air landings.

The structure of the aircraft was identical to that of An-10, but it had a lifted tail with a cut-down stern.

Accommodation: 6 persons (pilot, co-pilot, navigator, air engineer, air mechanic, radio operator).

The crew compartment was pressurized. The cargo compartment dimensions - 13.5x2x2.5 meters.

The aircraft is equipped with a crane-beam, transporter, and hoist which provide for loading and unloading of heavy and bulky industrial hardware.

Empty mass - 36,000 kg; take-off mass - up to 61,000 kg; cargo mass - up to 20,000 kg; cruising speed - 580 kph; landing speed - 215 kph; service ceiling - 10,000 meters; fully loaded range - 500 km, with 10-ton cargo - over 4,000km.

The An-12 can operate from grass strips. Take-off - 900 meters; landing run - 850 meters.

The maiden flight was performed by test-pilot Ya. Vernikov on December 10, 1957.

In flights performed by test-pilots G. Lysenko and I. Goncharov, crews took their places and different systems were tested. The tests were of paramount importance, not only for the An-12, but for all heavy turboprops. The tests foresaw one or even two engine failures. The tests resulted in developing a negative thrust auto-feathering unit which, in case of 600 kg negative thrust, put the failed engine propeller into auto-feathering position. The negative thrust auto-feathering device increased the flight safety of turboprop aircraft. The plane's fall into a spin was also tested. The fall into a spin of a heavy turboprop was first tested in our country.

For many years An-12 was the main aircraft in service with the Soviet Airborne troops.

The exhibited aircraft (No. 8900203) was constructed in March 1958.

The aircraft was production tested till September. The flight test time made over 32 hours. Then it was delivered to a unit where it remained in service till 1964.

The last flight was performed by the unit pilot Murashov on January 4, 1964.

In all, the flying time of the aircraft was 818 hours, and it performed 562 landings.

On January 10, 1964, the aircraft entered Service with Zhukovsky Academy training base, and on July 17, 1973, it became an exhibit of the AF Museum.

# An-24 Airliner

The development of the An-24 began in early 1958. It was to replace the aging An-2, Il-12, and Il-14 in inter-regional lines.

The twin-engined high-wing had a number of novelties, which became customary in aircraft constructing, e.g., wide use of binding and welding instead of traditional riveting.

The wing span of the aircraft is 29.2 meters; wing area - 75 square meters; length - 23.8 meters; take-off weight - 21,000 kg; number of seats - 50; cruising speed - 450 kph; range - 650-2,000 km (depending on load).

The maiden flight of An-24 was carried out by G. Lysenko's crew on October 20, 1959.

The aircraft is powered by two AI-24 turboprops, 2,550 kg HP thrust each. The modifications to the aircraft included an additional RU-19 900 kg thrust turbojet, designed by S. Tumansky. That turbojet, serving as a power-unit, provided the aircraft with electrical energy during parking and main engines starting. The aircraft can continue the take-off with one working engine and can also fly in that situation at an altitude of up to 3,000 meters.

Various modifications of the aircraft are in use with the national economy: transport, aerial photography, etc.

The exhibit (No. 47300903) was constructed in 1964. On June 1964 the aircraft was delivered to the Lvov Civil Aviation detachment where the routine flights began.

The flying time of the exhibit is 27,444 hours, and 24,962 landings were carried out.

The aircraft was delivered to the Museum from Lvov on May 24, 1979.

# An-22 Military Transport

The An-22 military transport, code named Antei, is the further development of the An-8 and An-12. It is an all-metal high-wing with nacelles of undercarriage closely attached to a large fuselage (all in the family of planes were called "broad-fuselaged"). The larger size required some structural changes. Thus, the single-fin vertical tail was replaced by a twin-fin tail. Such a modification proved to be highly viable: it was used in designing the An-24 "Ruslan" (150-ton load carrying capacity) and the An-225 "Mriya" (250-ton load).

The An-22 was designed for airlifting bulky loads up to 80,000 kg. The length of its cargo compartment is 33m; width - 4.4 meters, height - 4.5 meters. Undercarriage - 12 wheels. The aircraft length - 55.5 meters; wing span - 64.4 meters; height - 12.5 meters. Powerplant - four NK-12 N. Kuznetsov turboprops, 15,000 HP thrust each. Every engine has two coaxial 4-blade propellers of headwing rotation.

The maiden flight was carried out by test-pilot Yu. Kurlin's crew on February 27, 1965. In the course of tests the aircraft gained the cruising speed of 560 kph; range with full load - 5,000 km; take-off run - 1,460m; landing run - 1,000 meters.

Several world records have been set. On October 26, 1967, test-pilot I. Davidov established 15 world records in one flight. He took off with the payload of 100,444.6 kg to the altitude of 7,848 meters. That record was broken in 1983. On April 17, 1975, Antei with 400,000 kg load, covered the distance of 5,000 km and gained the cruising speed of 584 kph in average.

An-22 was twice demonstrated at the Paris Air Show and was one of their more popular sights. It had been developed and constructed to meet the requirements of Russia's industry. Bulky loads can be rapidly delivered to remote regions such as Siberia and the Far East. The aircraft can airlift equipment which cannot be transported by railroad. For example, when the gas industry of the North and Tyumen were in urgent need of two mobile gas-turbine power stations, the An-22 coped with the task in an hour. Ground transportation would have required several months.

The Museum exhibit flew to Monino piloted by Lieutenant-Colonel Bobrovsky-Belov's crew, on January 29, 1988.

# Be-12 Patrol ASW Amphibious Aircraft

In 1957 G. Beriev's Design Office developed an all-metal Be-12 amphibious aircraft. Two AI-20D turboprop engines, 5,180 kg HP each, attached to high wings provide speeds of up to 600 km/h. The fuel endurance is up to 7,500 km. Service ceiling - 10,500 km; take-off mass - 35,000 kg. Crew of 4 men.

The Be-12 can take off from and land on both the ground and sea surfaces.

The aircraft is fitted with powerful radar equipment, sonobuoy kit, and depth charges.

The maiden flight was performed on October 18, 1960.

The Be-12 amphibious plane was demonstrated at an air parade in Domodedovo.

The aircraft has set 42 records: speed, range, and weight-lifting ability.

The Be-12 is in service with the Navy aviation, being used for sea patrol missions. It is organic to the ASW forces of the country.

The production number of the exhibit is No 5600302.

In all, the aircraft has logged 809 flying hours, having performed 170 landings. It was delivered to the Museum on June 5, 1974, from an operational unit.

# Be-32 Airliner

By the late 1960s, the Beriev Design Office had developed the Be-30 for local airlines. Later it was modified and designated the Be-32. The latter was an all-metal monoplane with mechanized high wings.

The powerplant included two TVD-10 wing mounted turbo-props (V. Glushakov Office), 950 kg HP each.

The system had the so-called "cross" propeller transmission.

This feature made it possible (in case of engine failure) to transmit the revolutions of the remaining engine to both propellers to avoid the asymmetric thrust and harmful resistance of the failed engine propeller wind milling.

The passenger version accommodated 15 seats, while the ambulance version had 9 lying and 6 sitting patients.

The aircraft was fitted with navigational and radiotechnical equipment and anti-icing systems, thus providing round-the-clock flights under any weather conditions.

The aircraft take-off weight was 5,700 kg; cruising speed - 480 kph; tanked fuel endurance - up to 1,200 km.

The exhibit was delivered to the Museum in April 1983.

# 5

## Supersonic Jet Aircraft

## MiG-19 Fighter

In the early 1950s the Mikoyan Design Office started the development of a single-seat supersonic fighter, designated later as the MiG-19.

The first variant performed its maiden flight on June 16, 1951, piloted by G. Sedov. It was followed by SM-2 and the SM-3 through SM-9 variants. The latter was designated as MiG-19.

The MiG-19 was a mid-wing with a 55 degree sweep. The wing span - 9.0 meters, wing area 25 sq. meters, the aircraft length - 12.54 meters. Initially, the horizontal tail was made up of a stabilizer and elevator. But, later it was replaced by an all-moving controlled stabilizer because the elevator had been losing its effectiveness at high speeds. The flight control system included an irreversible hydraulic actuator which trimmed out control surfaces. To provide control system self-centering and habitual pilot's efforts, an artificial feel unit was employed.

In case of hydraulic system failure (the system provided hydraulic actuator functioning) the stabilizer control was performed by an electromechanism.

The ordinary mass of the aircraft was 7,560 kg, max - 8,630 kg.

The undercarriage was of usual type with nose wheel. To make the landing run shorter a brake parachute was used.

The first prototypes were powered by two AM-5 A. Mikulin turbojets, 2,000 kg HP each, attached to the aft fuselage. Later they were replaced by RD-9B engines with afterburner (3,300 kg thrust with afterburning). Powered by these engines the aircraft gained

supersonic speeds of up to Mach 1.4. The ceiling was 16,700-17,600 meters.

The aircraft was armed with three 30mm guns, two of them were mounted in the stem of the outer wing, and the third was in the front fuselage. Bombs and rockets were on wing pylons.

The aircraft went to the production line and entered service with the USSR and some other countries' Air Forces.

The MiG-19 was the country's first production jet aircraft which exceeded the Mach 1 speed.

It was constructed in Gorky and Novosibirsk.

The aircraft had several versions: front-line fighter (MiG-19S), interceptor (MiG-19P) and a modified interceptor (MiG-19PM). All of them differed in weapons, flight mass and performance (max speed varied from 1,250 to 1,450 kph).

The exhibited interceptor version was constructed in 1957. It is armed with four air-to-air missiles. It was operated by several flying units, having logged 964 hours.

The aircraft was delivered to the Museum on June 19, 1973.

# MiG-21 Fighter

Several variants of the MiG-21 (E-2, -4, -5, -6) proceeded the aircraft entering a large-scale production line. The sweep-wing E-2 was first flown by G. Mossolov on February 14, 1954, the delta-wing E-4 by G. Sedov on June 16, 1956. It was the E-5 version that V. Nefedov piloted which finally gained the planned speed of 2,000 kph at an altitude of 18,000 meters. During the seventh high-speed flight the engine failed because of a surge. The pilot was not able to restart the engine. He chose not to eject, instead making the decision to land. The landing was aggravated by a hydraulic system failure, and a delay in the electrical system switch-on. The flight resulted in an accident.

Having thoroughly analyzed the accident the experts found it necessary to modify the engine and flight control system. In 1958 the new E-6 powered by S. Tumanskiy R-11F-300 engines (5,600 kg HP thrust) went through complex test flights, and in 1959 the aircraft designated MiG-21 went to the production line.

The aircraft was constructed for many years. The production included various modifications: MiG-21F and MiG-21F13 front-line fighters, the MiG-21P, MiG-21PF, and MiG-21PFM interceptors, MiG-21R recon planes, MiG-21U trainers, and some others.

The MiG-21 was a delta mid-wing, the leading-edge sweep being 57 degrees. The fuselage was of oval cross section. An upper fuselage fairing stretched from the cabin to the fin. To improve the directional stability the aircraft had an under-fuselage strake. To make braking intensive in maneuvering there were three air-brake flaps; two on the sides of the fuselage and one under it. There was also a detection and aiming radar inside the intake cone.

Aircraft dimensions (MiG-21F13 version): wing span - 7,15 meters, wing area 22.95 square meters, length - 13.46 meters.

The first versions of the aircraft were powered by R-11F-300 turbojet (5,600 kg HP thrust), being replaced by R-13-300 (6,800 kg HP) and R-25-3— (7,300 kg HP) on the last production aircraft.

Dimensions (MiG-21PFM version): wing span - 7.154 meters, wing area - 22.95 square meters, length - 14.5 meters.

The aircraft weapons varied depending on the functions and could include air-to-air missiles, jet projectile pods, 30 mm guns, and bombs.

The MiG-21 of the first modifications was the world's lightest jet fighter. Its take off mass initially was 7,570 kg, then 8,950 kg (with external tanks - up to 9,080 kg). Max speed - over 2,200 kph, service ceiling - 19,000 meters, range without external tanks - 1,100 km with tanks - up to 1,900 km.

The MiG-21 of various modifications had set a number of records. In October 1959 the aircraft piloted by G. Mosolov gained the max speed of 2,500 kph and average speed of 2,388 kph at an altitude of 15-20 km, and in April 1961 the aircraft reached a zoom ceiling of 34,714 meters.

The MiG-21 was in service with the Armed Forces of the USSR and other countries. It was operated in the Vietnam War and in the Middle East conflicts.

The exhibited MiG-21PFS (No. 95210102) was constructed in June 1964. The last flight was performed by Rukhlyatko on October 29, 1970. In all, the aircraft had logged 284 hours. It was delivered to the Museum on June 30, 1971.

# MiG-21I ("Tu-144 Analogue")

In the 1960s the A. Tupolev Design Office began development on a new supersonic aircraft—the Tu-144, a tailless aircraft with a wing of ogival aerodynamic configuration. Once the work had started, it was decided to create its analogue—a laboratory aircraft—the MiG-21I.

Such an aircraft originated from the MiG-21. It was also a tailless plane with a wing analogous to that of the Tu-144, with the same arrangement of lateral-longitudinal control. Two prototypes were built. The span was 11.5 meters, length - 15 meters, take-off mass - 9,000 kg.

The test flights of the first prototype began on April 18, 1968, and were performed by test-pilots O. Gudkov and I. Volk. The test-flights made it possible to study the wing angle-of-attack charac-teristics, speeds, Mach numbers, stability, and handling of the supersonic airliner.

The first Tu-144 crew (E. Elyan, meters. Kozlov, S. Agapov) went through a training flight on the MiG-21I.

The second prototype of the MiG-21I (No. 010103) exhibited in the Museum was built at the A. Mikoyan Design Office factory in the early 1970s. The aircraft is powered by a R-11F2S-300 engine with 6,200 kg HP thrust.

On January 1970 test flights were begun by I. Volk. He was to test the speed and altitude range (up to 2,500 kph at up to 20,000 m) of the Tu-144. The last flight was carried out by I. Volk on August 14, 1979.

In all, the aircraft had logged 200 hours in 311 flights.

The exhibit was delivered to the Museum in 1980.

# MiG-23 Multi-Purpose Front-Line Fighter

In the mid-1960s the A. Mikoyan Design Office began development on the MiG-23 single-seat multi-purpose front-line fighter. The maiden flight took place on April 10, 1967. The aircraft was piloted by A. Fedotov.

The function of the aircraft was to fight air targets under any weather conditions by day and night in an electronic warfare environment. The aircraft was also designed for operating against visu-

ally detected ground targets and carrying out reconnaissance. To solve those problems, it was required to increase the aircraft speed envelope.

The MiG-23 is a variable-geometry high-wing aircraft. The outer wing has three fixed sweep angles: 16 degree - take off, landing, and loitering; 45 degree - maneuverable air combat, and aerobatics; 72 degrees - high-speed flying. The span varied from 14m to 7.8 meters, and wing area - from 37.35 square m to 34.1 square m respectively. The length of the aircraft is 15.75 meters.

To increase the flight stability at supersonic speeds, the aircraft is fitted with an additional foldable undercarriage fin. The MiG-23 structure contains some other novelties which proved to be effective in aircraft building.

To provide effective lateral control in full speed range, the horizontal tail unit is of a differentially deflected stabilizer surfaces arrangement. For longitudinal control both the stabilizer sides deflection is synchronized.

To provide heavy braking in a maneuverable air combat, lateral two-dimensional fully variable air intakes are attached to the fuselage.

The MiG-23 is powered by a turbojet by-pass engine with afterburner. The first production aircraft had R-27F2-300 engine (10,200 kg HP thrust), the succeeding ones - R-29F-300 (11,500 kg HP), later replaced by R-35F-300 (12,700 kg HP thrust).

Aircraft weapons: 6 short range air-to-air missiles, AAM anti-jamming missiles, jet projectiles, GSh-23 guns, bombs up to 2,000 kg. External armament mass - 3,000 kg.

Take-off mass - up to 16,000 kg.

IR systems provide air target interception at high and low altitudes in an electronic warfare environment.

The MiG-23 went through a large-scale test program, having reached the max speed of 2,500 kph, 18,000 m service ceiling, and 2,250 km range.

1969 saw the start of the series production run of the aircraft. It remained in service with the Soviet Air Force for over 20 years.

There are two MiG-23s in the Museum, demonstrating the max and minimum wing sweep.

The MiG-23 (No. 1) was constructed in 1967. The first test flight was performed by A. Fedotov. It was he who carried out the last flight on December 8, 1969. The aircraft had logged 178 flying hours. It was delivered to the Museum on June 30, 1971.

The MiG-23 (No. 3) was constructed on August 8, 1968.

The first test flight was performed by test-pilot Komarov on September 24, 1968. It was he who performed the last flight on September 20, 1973. All in all the aircraft had logged 225 hours. It was delivered to the Muscum on December 12, 1973.

# MiG-25 Multi-Purpose Fighter

The advent of the MiG-25, developed by A. Mikoyan Design Office, was a remarkable achievement in the speed of atmospheric flying vehicles. MiG-25 is a tapered high-wing of low relative thickness. A broad fuselage with lift qualities has two-dimensional air intakes on both sides. The vertical tail is of the twin-fin type. The horizontal tail is a differentially controlled stabilizer, providing high effectiveness for both longitudinal and lateral control in full speed range.

Wing Span is 14.1 meters, wing area - 61.9m, aircraft length - 19.75 meters.

The kinetic heating of the structure at supersonic speeds required heat-resistant materials.

The MiG-25 maiden flight was carried out in March 1964. The aircraft was piloted by A. Fedotov.

The aircraft is powered by two R-15BD300 turbojets (designer S. Tumansky), each 11,200 kg HP thrust. These engines make it possible to operate at speeds up to 3,000 kph and to perform energetic maneuvers in 2,500-3,000 kph speed range. The take-off mass of the aircraft is 37,000 kg, service ceiling exceeds 22,000 meters, the absolute ceiling is over 37,000 ml range - 1,750 km. To gain 20,000 m altitude, 6-8 minutes are needed ( the record was 35,000 m in 4 minutes 11 seconds).

The MiG-25 has the following versions: reconnaissance aircraft, strike aircraft, and interceptor.

The armament depends on the aircraft missions and may include AAMs, bombs, and guns.

To develop descent and landing of the Buran versatile spacecraft and its escort during landing, an airborne laboratory was constructed on the base of the MiG-25.

The exhibited MiG-25 (No. 0200001) was constructed on December 31, 1966. It was first flown by test-pilot L. Mininko on March 30, 1967. The last flight was performed by A. Fedotov on June 8, 1973.

The aircraft had logged 197 flying hours. On August 23, 1973, it was delivered to Zhukovsky Engineering Academy training base and in September 1979 it became an exhibit of the Air Force Museum.

# MiG-29 Front-Line Fighter

In 1977 the A. Mikoyan Office, headed by Chief Designer R. Belyakov, developed the MiG-29 fourth-generation front-line fighter. It was designed to fight enemy air-power, to protect troops and front rear objectives, and to counter enemy air reconnaissance.

The latest achievements in science and technology were used in designing the plane. The airframe was of an integral aerodynamic configuration with not only wing high lift, but that of the fuselage as well. To make the plane lighter, composite materials were used. The aircraft has a high thrust-to-weight ratio with engine thrust power exceeding force of gravity at altitudes up to 3,000 meters. That provides high maneuverability of the aircraft both in horizontal and vertical planes.

Two RD-33 engines (V. Klimov Design Office) with 8,340 kg HP thrust each allow for short-range maneuverable dogfights with up to G9 loads. High thrust-to-weight ratio and good aerodynamic features provide high characteristics of take-off run and acceleration, low turn radius, high turn angle speeds. Start thrust-to-weight ratio of MiG-29 is over 1.1.

The wing span is 11.36 meters, wing area - 35.2 square meters, length - 17.3 meters.

Armament : 30 mm GSh-30 gun, AAM, unguided projectiles and bombs. The aircraft has 8 storage pylons.

The aircraft is fitted with a radar and detection/tracking optical electronic system. The complex includes helmet-mounted designation.

To counter enemy airborne and ground-based weapons, the MiG-29 has a defensive system.

The maiden flight was performed by A. Fedotov on October 6, 1977. Then the aircraft went through production and state tests. In 1982 it entered the production line and went into service with the Air Force. It was also delivered to some foreign countries. One Luftwaffe regiment is equipped with the MiG-29.

During test flights the max. speed of over 2,450 kph was achieved, while minimum speed was 180 kph. The service ceiling - 18,000 meters, range - 2,100 km, ground rate of climb was 330 mps with 15,240 kg take-off mass.

The MiG-29 was demonstrated in 1989 at the Farnborough Air Show (held in France and Canada), and there it was recognized as the world's best front-line fighter.

On November 2, 1989, the MiG-29 piloted by Aubakirov landed on the deck of the aircraft-carrier cruiser "Admiral Kuznetsov." Later the aircraft successfully carried out a jump ship-board take-off followed by landing with the use of an arrester system. So, the aircraft proved it's capability as a carrier-based plane.

The aircraft has been on the production line in Nizhny Novgorod and Moscow since 1982.

The exhibit was delivered to the Museum on November 11, 1986.

# Buran Analogue Prototype

The vehicle is a replica of the Buran military spacecraft and is designed for testing the latter at the atmospheric landing stage and on landing. Several take-off runs and hops have been performed with independent wheel-undercarriage take-off. The vehicle was piloted by A. Fedotov, V. Menitsky, and I. Volk.

On November 28, 1975, test-pilot Fastovets performed the first 4-minute flight. In subsequent flights the analogue was dropped from the airborne laboratory at 5,500 m and landed on a grass airfield. Those flights were carried out by test-pilots Fedotov, Fastovets, Ostapenko, Volk, and Uryadov in 1976-78.

In all, the vehicle had logged 32 hours 45 minutes and was delivered to the Museum on September 1, 1989.

# Su-7 Front-Line Fighter

The Su-7 (P. Sukhoi Design Office) was the Su series' second aircraft: in 1945 test-pilot G. Komarov tested the piston-engine Su-7 with RD-1 jet booster, but the test flights were interrupted because of booster limitations. 10 years later the same designation was given to a new machine which proved to be very good.

The Su-7 one-seater is a mid-wing with 55 degree sweep. The horizontal tail is an all-moving stabilizer. For the first time in the country's aircraft construction a variable air intake and KS-1 ejection seat were used.

The wing span of the aircraft was 9.3 meters, wing area - 34.0 sq.m, length - 18.05 meters, take-off mass - 13,600 kg.

The fixed armament consisted of two NR-30 guns in the wing stem. The four underwing and two underfuselage pylons could carry AAMs and ASMs, as well as unguided missile pods.

The aircraft was powered by the Al-7F-1 turbojet engine with 7,800 kg HP thrust, which practically was still under development. To a certain extent it was a risk, but there was no way out: only a A. Lyulka engine could provide the required characteristic—a speed of 1,800 kph.

On September 8, 1955, during a high-speed taxi run, the aircraft suddenly went airborne. Flight termination was impossible and the machine took to the air. Unavoidably, test-pilot A. Kochetkov performed the first flight of Su-7. Later the flight tests were carried out by V. Makhalin.

In the course of test flights the speed of 2,170 kph was achieved, twice exceeding the sonic speed. Such a speed was reached in the USSR for the first time.

The service ceiling was 19,000 meters.

The aircraft went to the production line in mid-1956.

The Su-7 was widely operated by the AF units. Later there appeared aircraft modifications, such as: Su-7B fighter-bomber, and Su-9 interceptor.

The exhibited Su-7 (No. 1707) was built in 1960. All in all it had logged 13 flying hours in a military unit.

On January 29, 1962, it was delivered to the Zhukovskiy Engineering Academy training base. On November 2, 1968, it became the Air Force Museum Exhibit.

# Su-9 Interceptor Aircraft

This aircraft had its predecessor. In 1946 the P. Sukhoi Design Office developed the firm's first jet aircraft. It was the Su-9 fighter powered by two RD-10 engines. The aircraft was tested by A. Kochetkov and had good characteristics for the time: speed up to 900 kph, ceiling 12,750 meters, range 1,100 km. But the fighter did not go to the production line because of the MiG-9 and Yak-15 being produced. For unexplained reasons Sukhoi decided to reuse the SU-9 designation for this second, totally new aircraft.

The development of the new supersonic interceptor, Su-9, began in 1953.

To minimize the development time, the aircraft structure practically originated from the Su-7. Rear fuselage, tail unit, on-board system, canopy, survival aids, and undercarriage wheels were the same. The engine, A1-7F-1, was also the same, with 7,800 kg HP thrust. The wing was of a delta type with pressurized torsion fuel tanks. The wing sweep was 60 degrees, span - 8.54m, aircraft length - 16.77 meters.

The Su-9 armament included 4 AAMs and Almaz radar.

On May 26, 1956, the first test flight was performed by V. Makhalin. In the course of the tests the aircraft reached the speed of 2,100 kph, ceiling - 20,100 meters, range 1,100 km.

The all-weather Su-9 interceptor went to the production line and entered service with the Air Defense air arm.

On June 24, 1956, an air show took place in Tushino. Six aircraft of new type were demonstrated there, the Su-7 and Su-9 among them.

The exhibited Su-9 (No. 0615308) was constructed in August 1959.

On September 24, 1959, the aircraft became operational.

All in all the aircraft had logged 30 hours 25 minutes in 52 sorties. The last flight of the aircraft was performed by pilot Garov, and it landed on Monino airfield on September 15, 1960. The plane was delivered to the Zhukovsky training base.

On September 1, 1969, the aircraft became the exhibit of the AF Museum.

# Su-7B Fighter-Bomber

In the late 1950s it became necessary to develop a new aircraft with the features of a bomber, an attack aircraft, and a fighter. The Su-7B, developed by the P. Sukhoi Office, became such a machine.

The design foresaw the following capabilities of the aircraft: toss bombing, aerobatics, and strikes against ground targets.

The aircraft originated from the Su-7 front-line fighter structure, which reduced the design time significantly.

The armament included two 30mm guns, two UB 16 unguided rocket pods, bombload up to 1,000kg, and air-to-surface unguided missiles.

The aircraft was powered by one Al-7F-1 engine with 7,800 kg HP thrust.

In April 1959 the production tests were begun by E. Solovyev. In the course of the tests the machine reached max speed of 2,230 kph, service ceiling - 20,100 meters, range - 1,100 km.

In 1960 the decision was made to begin production of the aircraft. Soon the first planes were delivered to operational units. The machine became one of the main aircraft for the Soviet Air Force.

Later the P. Sukhoy Office developed several modifications capable of operating not only from concrete runways which are highly vulnerable in war time, but also from grass strips.

The museum has two versions of the Su-7B with ski and wheel-ski landing gears.

The production Su-7 (No. 3608) was constructed in 1961, went through all types of tests, was fitted with ski landing gear and designated the Su-7L. The first flight with ski landing gear was carried out by I. Ryabchikov.

The machine was flown for the last time by pilot Poltornov on February 22,1971. In all, it had logged 75 flying hours.

The exhibited Su-7L tests revealed that the ski landing gear could be used (in the absence of snow) only on very humid ground. This resulted in the creation of a wheel-ski gear: the small ski was near the wheel. In taking off from hard ground or a concrete runway, the ski did not touch the surface, instead remaining slightly raised over the ground. If the ground is soft, the wheel sinks deeper

into it, activating the ski, which prevents the wheel from a come-down.

The exhibited Su-7Bkl (No. 5812) was built in October 1965.

It was operated by a military unit from February 11 till April 26, 1966. On that day the last flight took place, performed by pilot Bereiko.

After a ground accident the aircraft was repaired and used as a ground trainer.

In December 1975 the aircraft was delivered to the Zhukovsky Engineering Academy training base and the same month it became an exhibit.

# Su-11 Fighter-Interceptor

The first aircraft of this designation was constructed shortly After World War II, a memorable event in Russian aviation.

It was the first aircraft powered by original home-made TR-1 A. Lyulka jets, a new family of the Design Office engines.

The first flight was performed by G. Shiyanov on May 28, 1947.

The aircraft went through test flying, and participated in the Tushino Air Parade (1947), but did not go to the production line.

The "new" Su-11 was a supersonic interceptor armed with two powerful AAMs and powered by the Al-7F-2-200 engine with 9,600 kg HP thrust. Crew - one pilot. Take-off mass - 14,200 kg. Wing span -8.54 meters. wing sweep - 60 degrees, aircraft length - 17.55 meters.

The first flight was performed by V. Ilyushin on February 21, 1958. Max speed - 2,200 kph, service ceiling - 20,100 meters.

The Su-11 (No. 0115307), exhibited in the Museum, was constructed in 1962. On August 13, the factory test-pilot Vilomov performed a flight.

On April 1963 the machine was delivered to an operational unit.

In all, the aircraft had logged 199 flying hours in 283 sorties.

On July 13, 1973, the plane was delivered to the Zhukovsky Engineering Academy training base, and on July 17 it became a Museum exhibit.

# Su-15 Fighter-Interceptor

The Su-15 supersonic interceptor was designed in 1963 to meet the requirements of long-range radar detection of the enemy (not to be confused with the Su-15 subsonic jet fighter designed by P. Sukhoi in 1949. The plane piloted by S. Anokhin had met with an accident as a result of vibrations and the decision was made not to restore it).

The Su-9 and Su-11 interceptors were fitted with a radar in the air inlet cone. To increase the interceptor's versatility, the radar was larger and could not be of previous accommodation. The avionics was accommodated in the nose fuselage and the inlets on the sides of the fuselage. So Su-15 one-seater had side intakes and delta wings with varying sweep of leading edge.

The wing area - 35.7 square meters, span - 10.53 meters, aircraft length - 20.5 meters.

The machine is powered by two R-11FAS-300 engines, 6,200 kg HP thrust each. The take-off mass - 17,900 kg.

The first flight was performed by V. Ilyushin on May 30, 1962.

In the course of the test flights the aircraft reached a max speed of 2,100 kph, service ceiling - 18,000m. The plane was equipped with four AAM guided missiles, the bomb load was up to 1,000 kg.

The exhibited Su-15 (No. 758) was manufactured in 1963.

On May 4, 1963 V. Ilyushin carried out the post-production test

flight and the aircraft was delivered to another organization for further tests.

During the test flights the aircraft had logged 209 flying hours.

On January 22, 1974, the aircraft was delivered to the Museum by ground transport.

# Su-17M Fighter-Bomber

In 1965 the P. Sukhoi Design Office developed the Su-17M, a variable sweep wing aircraft. The concept of such machines appeared because of conflicting requirements for the wing, depending on various flying modes. Flights at great speeds require highly swept wings, but take-off, landing, and long-range flights require a minimum sweep. With the increase of maximum speeds the conflict sharpened.

To reduce the design stage and simplify the construction, the P. Sukhoi Office developed the Su-17M on the base of the Su-7B production plane.

The wing mounted outer wing turn unit and the sweep control mechanism led to a take-off mass increase of up to 19,800 kg and to the reduction of wing fuel. That could have resulted in a worsening of performance.

But the first flight performed by V. Ilyushin on August 2, 1966, and further tests proved the Su-17M to be a high-performance aircraft. The range extended: with two external tanks it made 2,800

km. The take-off and landing speeds were reduced, but the max speed was 2,250 kph. The landing and take-off runs decreased, so a shorter runway was required.

The aircraft is fitted with an Al-21F-3 turbojet, with 11,200 kg HP afterburning thrust.

The Su-17M has versatile firepower. The bomb load is 4,000 kg. The aircraft is armed with guided and unguided ASMs and AAMs, and two 30-mm guns. The aircraft was produced in several versions and is in service with the fighter-bomber aviation.

The exhibited Su-17M-3 (No. 22301) was constructed in November 1976.

From January 1977 till October 1979 the machine was operated, having logged 200 flying hours. The last flight was performed by V. Ilyushin on October 25, 1979.

It was delivered to the Museum from the Zhukovsky Engineering Academy training base in October 1989.

# Su-100 Prototype Bomber

In the early 1960s Chief Designer P. Sukhoi and N. Chernyakhov, along with their colleagues, began the competitive development of a new prototype supersonic carrier-bomber—the Su-100 (factory designation - T-4).

The assigned task was highly complicated. It was necessary to reach the max speed of 3,200 kph and a cruising speed of 3,000 kph and to provide a long-duration flight at these speeds. So, the aircraft was exposed not to the sonic, but to the thermal barrier.

The Su-100 range was to be 6,000 km with 114,000 kg take-off mass, and 7,000 km with max take-off weight of 135,000 kg.

The structure of the aircraft contained many new and extraordinary features.

The crew cabin is not projected but the nose of the 45 m fuselage is of a down deflected arrangement. These provide the downward view.

The delta wing is thin and highly swept along the sharp leading edge. The airframe is welded from titanium alloy and high strength stainless steel, as in flight the temperature of the skin could reach up to 300 degrees C. The aircraft is powered by four RD-36-41 turbojets (P. Kolesov Design Office) with 16,000 kg HP after-burning thrust and 10,000 kg HP nominal thrust each. They are on an underfuselage arrangement—two in one duct.

The aircraft is equipped with electric remote plane and engine control systems of higher reliability, with fourfold reservation of every control channel.

The avionics system included several complexes: navigational aiming—on the base of a long-range, forward-looking surveillance radar—and a reconnaissance complex with optical and IR sensors.

On August 22, 1972, V. Ilyushin took the aircraft into the air. Ten test flights were performed and the speed gained was Mach 1.7. But, in 1975, the tests and further development of the aircraft were ceased most probably because of technological difficulties of the Su-100 production.

To a certain extent the Su-100 had outstripped its time. Many technological concepts of its development, however, continue to be used in designing flying vehicles of the modern generation.

The exhibited aircraft was delivered to the AF Museum in 1982.

ce
ra
th
19

sq

in
pc
Su
le
fl
at
at

# Su-24 Front Bomber

The Air Defenses improvement increased bombers' vulnerability at high speeds and altitudes as well as at respectively low speeds at ground level. Such a front bomber capable of supersonic low-level missions was badly needed. The aircraft was developed by the P. O. Sukhoi Design Office. It combined front bomber and attack aircraft qualities. It was capable of:

- performing supersonic as well as subsonic flights conducting day/night all-weather operations within full altitude range;

- precision destruction of ground and sea targets in manual and automatic control modes;

- detecting and destroying transport, liaison, adjustment, and other type aircraft and helicopters day and night under good weather conditions with IR AAMs and under clear visibility with guns.

The Su-24 two-seat front bomber is a high-wing with a monocoque fuselage.

Accommodation: pilot and navigator side-by-side in a pressurized cabin with K-36D ejection seats.

The aircraft is equipped with a multimode radar, automatic flight control system, and an integral navigation/attack complex, enabling it to fly an assigned route in the autonomous navigation mode.

The aircraft armament is mounted on 8 stores and comprises: 100-1,500 kg of bombs and bomb clusters; IR air-to-air missiles; laser- and television guided missiles; 57-370 mm rockets; built in 23-mm gun; three removable revolving cannons (ammunition store of 400 shells each in special pods).

The Su-24 was first flown by test-pilot V. Ilyushin on June 17, 1970.

onds to perform a 360 degree steep turn. From 1986-1988 the aircraft set 27 rate of climb and level flight altitude world records.

In November 1989, for the first time in our country, test-pilot V. Pugachev took off and landed the Su-27 from the "Admiral Kuznetsov" aircraft carrier.

The Su-27 was demonstrated at the international aviation show in Paris, Singapore, and other countries. V. Pugachev performed the "Cobra" flight maneuver while flying the aircraft. The Su-27 is one of the world's best fighter-interceptors due to the expedient combination of aerodynamic configuration, high thrust-to-weight ratio, and powerful armament.

The aircraft, exhibited in the Museum, was delivered there from the Sukhoi Design Office on January 23, 1986.

# La-250 Fighter-Interceptor

In 1956 the Lavochkin Design Office developed the La-250 two-seat all-weather fighter-interceptor. It was one of the first delta-wing variable-incidence tailplane aircraft. It was an all-metal structure.

The aircraft control system is comprised of irreversible two-chamber hydraulic actuators powered by two hydraulic systems.

Two Al-7F turbojet engines with 6,500 kg thrust each are installed on the sides of the extremely long fuselage. The air intakes are separated from the fuselage.

Internal fuel tank capacity is 8,000 kg, one external tank fuel capacity is 1,100 kg.

The armament included a radar sight installed on the aircraft nose and two powerful air-to air-missiles.

Four aircraft of this type were built, three of which were tested in 1956-1958. Test-pilots A. Kochetkov, N. Zakharov and A. Brodsky took part in the test program. A number of complications during the flight test occurred. The first aircraft crashed because of engine failure and undercarriage collapse (a production defect) while landing. As a result the test program dragged on and, because of Lavochkin's work on pilotless weapon systems and, unfortunately, on June 9, 1960, his death finally forced the termination of testing on the aircraft.

Nevertheless, a number of successful flights were performed during the test program which proved the following characteristics: maximum speed - 1,980 km/hr, service ceiling - 18,500 meters; flight range - 2,800 km; flight endurance - 2.5 hours; take-off weight - 25,000 kg.

The La-250 (No. 125000) exhibited in the Museum was produced in November 1958. It was not tested, but it was used as a training device at the Zhukovsky Air Force Engineering Academy. In 1967 it was delivered to the Museum.

# Yak-27R Reconnaissance Aircraft

The Yak-27R supersonic all-weather two-engine two-seat reconnaissance aircraft was developed by the Yakovlev Design Office in 1958. It is the Yak-25 family successor: it is a mid-wing monoplane with a swept wing and bicycle undercarriage.

The aircraft is powered by two RD-9F turbojet engines, 3,660 kg thrust each, installed on the wing. Take-off weight is 10,700 kg. After the evaluation and test program it was adopted for service. During the flight tests the aircraft revealed the following characteristics: maximum speed at 11,000 m - 1,285 km/h, service ceiling - 16,500 meters; range of flight - 2,210 km.

Armament: one NR-23 gun, APA-C and AKAFU-3 photo cameras.

The Yak-27 exhibited in the Museum was produced in November 1959.

On January 26, 1960, it was delivered to an air unit. The aircraft was flown for the first time by pilot Lysenko on May 24, 1963. The aircraft flight time is 420 hours, and it performed 497 landings. It was delivered to the Museum on July 30, 1972.

# Yak-28L Front Bomber

The development of the Yak-28 supersonic front bomber began in the Yakovlev Design Office in late 1958. It's a mid-wing monoplane with a swept wing and bicycle undercarriage.

The aircraft is powered by two turbojet engines, 5,900 kg HP thrust each, installed on the wing.

Dimensions: wing span - 11.64 meters, wing area - 35.25 square meters, fuselage length - 20.34 meters. Armament: one two-barrel 23-mm gun; 1,000-3,000 kg bombload carried internally; "Initiative" and "Lotos" sighting systems. As a result the aircraft was designated Yak-28I and Yak-28L respectively. Weights: take-off weight (in different variants) - 15,545-17,465 kg. Accommodation: a pilot and a navigator. While being tested the aircraft revealed the following characteristics: maximum speed - 2,060 km/h, service ceiling - 16,700 meters; range of flight with 2x1000 kg external fuel tanks - 3,740 km.

After the evaluation tests the aircraft was adopted for service and entered serial production.

The Yak-28 modifications were Yak-28R provided with photo and radar equipment and Yak-28P fighter-interceptor armed with a radar sight and air-to-air missiles.

The Yak-28L (No. 2920902) was produced in July, 1962. The last flight was performed by pilot Lvov on January 28, 1966. The aircraft flight time is 245 hours. It made 258 landings. In July 1966 it was delivered to the Zhukovsky Air Force Engineering Academy training base, and it was passed over to the Museum on November 1975.

# Tu-22 Bomber

The Tu-22 is a low-wing monoplane. It is powered by two VD-7M double-flow turbojet engines, 16,000 kg HP thrust each with after-burners.

Dimensions: wing span - 23.6 meters; sweepback at leading-edge - 56 degrees; fuselage length - 41.6 meters, take-off weight - 92 tons.

Accommodation: three-man crew. Performance: maximum speed - 1,640 km/h, landing speed - 290 km/h, take-off run 2,800 meters, landing run -1,850 meters, service ceiling - 13,500 meters, range of flight - 5,600km.

Armament: 9 tons bombload; one 23-mm gun. The aircraft was first flown by test-pilot Yu. Alasheyev on June 22, 1958. The further flights were performed by the same pilot.

The Tu-22 was adopted for service in 1965. The Tu-22 (No. 505005) exhibited in the Museum was produced in 1960. After flight delivery tests it was acquired by an organization. It was flown there until September 28, 1962. The aircraft flight time was 98 hours.

On September 28, 1962, the aircraft was delivered to an air unit. It was operational till October 18, 1965. The aircraft's last flight was made by pilot Cherno-Ivanov on October 27, 1965: it flew to the Zhukovsky Air Force Engineering Academy training base. On February 18, 1977, it was exhibited in the Air Force Museum. The aircraft total flight time is 350 hours.

# Tu-22M Missile Carrier

The Tu-22M ("Backfire" under the NATO designation) is a low-set variable-sweep wing missile carrier. Wing sweep is 20-60 degrees, and the wingspan is 24 meters. Fuselage length is 400 meters, and the weight is 122,000 kg.

The aircraft possesses two NK-20 engines 25,000 kg thrust each. It had a four-man crew.

Performance: maximum speed is over 2,000 km/h. Combat radius is 2,200 km. Service ceiling is 18,000 meters. It can be refueled in flight.

Armament: three Kh-22 air-to-surface missiles and 24,000 kg bombs.

Self-defense weapons: one 23-mm twin-barrel cannon installed in the tail, air-to-air forward hemisphere missiles and passive ECM. The aircraft is equipped with satellite navigation and automatic control systems (terrain-following mode including).

The aircraft was first flown by test-pilot V. Borisov on August 10, 1969. The aircraft was put into serial production and adopted for service in the Soviet Air Force.

The Tu-22M (type 45 No 5019029) exhibited in the Museum was produced in 1970. It was tested on March 23, 1970, through January 15, 1971. In October 1971 the aircraft engines were replaced by more powerful ones at the plant. Simultaneously, the aircraft fuselage was strengthened and it was provided with a brake parachute.

On March 13, 1974, the aircraft was subjected to 2-4 multiple overloadings three times. As a result further employment became possible only on the ground for training purposes.

On February 25, 1975, pilot Borisov flew the aircraft for the last time. It landed on the Monino airfield and was delivered to the Zhukovsky Air Force Engineering Academy training base. The aircraft became and exhibit in October 1989.

# Tu-128 Interceptor

The Tu-128 all-weather supersonic interceptor was produced in 1960. The aircraft is powered by two Al-7F-2 (A. Lyulka design) engines 10,100 kg thrust each with after-burning. The crew is two pilots. Weight: 43,600 kg. Armament: four R-4 air-to-air missiles. On March 18, 1961 pilot meters. Kozlov began testing the aircraft, which lasted till May 6, 1972.

The aircraft demonstrated the following characteristics: maximum speed at 12,000 - 1,600 km/h. maximum speed at ground-level - 850 km/h, service ceiling - 15,600 meters. Range of flight with four missiles - 2,000 km. The Tu-128 exhibited in the Museum is a prototype version.

The aircraft flight time is 398 hours 35 minutes. On February 2, 1973 the aircraft was delivered to the Museum as its flying life had come to an end.

# Tu-144 Airliner

Creating a supersonic airliner required solving a great number of scientific and technological problems.

It was necessary to find a specially shaped aircraft that would almost twice exceed the aerodynamic qualities of any existing military supersonic planes. It was also necessary to solve the problems of stability and balance in subsonic, transonic, and supersonic areas, and to create a control system that would correspond to the aircraft and such multimode versatility. It was necessary to obtain heat-proof structural materials, lubricants, air-tightness, as well as structure types capable of functioning under aerodynamically heat-

intensive conditions for a long period of time; to design powerful engines providing high affectivity in a supersonic flight; and to create highly effective air inlets operational within wide altitude and speed ranges.

France and Great Britain had been cooperating in developing the "Concord" supersonic airliner by that time.

Tupolev's Design Office constructed the Tu-144 plane. The aircraft is a low-wing monoplane with an ogival delta-wing; span 28.8 m with a variable sweep angle. There are elevons used as aile-

rons and an elevator on the back side of the wing. Aircraft length is 65.5 meters, with a crew of 3 men.

While taking off and landing the nose of the fuselage deviated 20 degrees downwards.

The aircraft was powered by four turbojet engines.

Two compartments accommodated 120 passengers. Maximum speed was expected to reach 2,500 km/h; range of flight - 6,500 km, ceiling - 18,000 meters; landing speed - 266 km/h; maximum take-off weight - 180,000 kg, payload - 12,000 kg. Fuel endurance - 95,000 kg.

The Tu-144's first flight was preceded by testing the control system on simulators as well as by flying an analogue MiG-21 that had a reduced wing fully identical to that of the Tu-144.

On December 31, 1968, pilots A. Yelyan and M. Kokhlov flew the Tu-144, the world's first supersonic passenger airliner.

Originally it was powered by NK-25 (N. Kuznetsov) engines, with 25 tons thrust and afterburner.

In June 1969 the Tu-144 exceeded M1 for the first time. On July 15, the same year, the airliner developed the speed of 2.443 km/h.

Nevertheless, further development of the airliner was not so easy. The NK-25 engine fuel consumption provided 3,500 km range of flight. The decision was made to install P. Kolesov engines at 25,000 kg thrust each without afterburning, which proved more effective and allowed the aircraft to reach a 6,500 km range of flight.

The world energy crisis and the consequent sharp increase in fuel prices, however, seriously undermined the profitability of supersonic passenger airliners.

At the same time, increased ecology requirements limited the possibility of such aircraft flights.

On June 3, 1973, the Tu-144 was performing a demonstration flight in Le Bourget, Paris, and crashed in an air accident.

Nevertheless, the Tu-144 design work continued. On November 30, 1974, the Tu-144 powered by Kolesov engines made a Moscow-Khabarovsk flight.

On December 26, 1975, there began Moscow-Alma-Ata operational test flights with cargoes and mail on board.

The Tu-144 first passenger flight was performed by a crew of Pilot Kuznetsov en route from Moscow to Alma-Ata. Regular flights were made until May 1978. They were ceased because of an air accident on May 28, 1978, when three people perished. The aircraft was withdrawn from service.

Nevertheless, the Tu-144 creation and operation experience as well as that of the Concord airliner made an important contribution to aviation science and was used in future aircraft building.

The Tu-144 (No. 10041) was produced at the Moscow "Opyt" (Experiment) machine-building factory in 1975. From October 1975 till February 1980 the aircraft was undergoing regular tests. Its flight time totals 582 hours 36 minutes. The aircraft made 320 landings.

The last flight was performed by test-pilot G. Voronchenko on February 29, 1980. The aircraft landed on Monino airfield and thus was delivered to the Air Force Museum.

# M-50 Strategic Missile Carrier

The M-20 was designed by the Myasischev Design Office in 1959.

Dimensions: length - 57 meters, delta wing span - 37 meters, height - 12 meters. Armament: two nuclear missiles (30 tons total weight) installed in a 15-m bay. No defensive weapons.

It was the first aircraft provided with electronic automatic controls: navigation and sighting systems; center-of-gravity shift system; power plant operational control system; and stability and controllability support system.

The M-50 was also the first aircraft provided with the fully automated remote control system (without a mechanic stand-by system). This enabled a two-man crew previously believed to be too small for a 200-ton strategic bomber.

Undercarriage: Bicycle landing gear with a ramping front-wheel leg and two legs installed on the tips of the outboard wings.

The M-50 production was based on some principal technological novelties. A new technological process of building the wing and the fuselage out of personal panels was developed. The load carrying elements and covering were made as an integral whole.

The internal engines were installed on pylons, while the external ones were mounted edge-to-edge with the wing tips.

Four P. Zubets Design Office RD-16-17 engines 18,000 kg thrust each were planned to be installed on the aircraft. But the production of these engines was delayed and the decision was made to use two VD-7A engines 11,000 kg thrust each temporarily and to mount them edge-to-edge to the wing tips and two VD-7A engines 16,000 kg thrust each mounted on the centerline wing pylons.

In October, 1959 test-pilots N. Goryainov and A. Lipko performed the first flight which was the beginning of the evaluation test program.

The maximum speed with the Zubets engines was expected to be 1.8, (i.e. about 2,000 km/h) with a range of flight over 6,000 km. But, contrary to expectations, the M-50 powered by other engines reached the level of sonic speed and developed speeds of more than Mach 1. Take off and landing characteristics proved higher than design data. All that allowed the designers to hope the design data would be obtained.

However, in the late 1950s the Strategic Rocket Forces were believed to be capable of providing the country's defense potential themselves without the other Fighting Services' participation. Such views could not exist for long, but it was quite enough for the Myasishchev Design Office to be disbanded and the M-50 advanced strategic bomber test program to be stopped.

The M-50 exhibited in the Museum was delivered from the V. Myasishchev experimental machine-building factory on October 30, 1968. There are neither systems and equipment nor engines on it.

# 6

## Helicopters

## Mi-1 Helicopter

Design of the GM-1 single three-seat helicopter (later designated Mi-1) began towards the end of 1947. The helicopter was provided with a three-bladed main rotor 14.3 m in diameter. Powerplant: seven-cylinder piston AI-26V aircooled engine, 575 HP.

A special engine fan was used for aircooling the engines. The helicopter cockpit accommodated a pilot and two passengers. Three prototype versions were built in succession. The first one was tested by test-pilot meters. Baikalov on September 29, 1948. But, during one of the flights, Baikalov was forced to leave the aircraft because of stability loss at 5,200 meters.

It was in 1954 when the helicopter first appeared in the North Pole.

The Mi-4 set a number of world records. The 2,000 kg cargo was lifted to 6,017 m altitude. It was widely employed in the USSR and abroad, having proved better than analogous foreign machines during competitive tests.

The Mi-4 (No. 1104) helicopter exhibited in the Museum was manufactured in April 1953. The helicopter was operational in air units. It flew 1,386 hours. It was delivered to the Museum on January 12, 1965.

# Mi-6 Helicopter

The development of the Mi-6 heavy assault helicopter began in the Mil Design Office in June 1954. In June 1957 it made its first flight.

The helicopter was powered by two D-28V Solovyov design semi-turbine engines with 11,000 HP general take-off power. The engine power was transmitted to the five-blade rotor (35 m in di-ameter) through the main reducer, and to the tail rotor through the intermediate and tail reducers. The helicopter power-to-weight ratio ensured a horizontal flight with one engine failure.

To discharge the rotor at high speeds the helicopter was provided with a 35 sq. m controllable wing carrying 25 per cent of the flight mass of the helicopter.

The fuselage accommodated the cockpit and the cargo compartment, and at 11.6 m long, 2.3 meters. wide, and 2.6 m high, enabled the planes to air-lift a 12,000 kg load internally and 8,000 kg externally. Loading and unloading was performed through the cargo hatch over the loading ramp in the rear fuselage. The assault version accommodated 41 lying wounded or sick and 2 nurses.

Normal take-off weight - 40,000 kg, maximum take-off weight - 42,000 kg; the total internal and external fuel endurance - 9,805 kg; the internal one - 6,316 kg.

Autopilot, complete navigation set, and anti-icer, as well as the presence of a navigator on board the helicopter, ensures the Mi-6's all-weather capability.

On June 5, 1957, the helicopter was first taken to the air by test-pilot R. Kaprelyan. In October of that same year the helicopter lifted 12,000 kg load to 2,432 m altitude. The helicopter revealed the following characteristics: maximum speed and altitude with normal take-off mass - 300 km/h and 4,5000 m respectively; range of flight with full fuel tanks and 4,500 kg load at 250 km/hr cruising speed is 1,040 km; flight endurance at 160-180 km/hr cruising speed is 2 hours 50 minutes.

In late 1959 the helicopter entered serial production, but its improvement continued. For its service life to be extended, the rotor blade construction was constantly being improved.

By 1970 the blade service life had improved up to 1,000 hours.

The Mi-6 helicopter set 16 world records. In 1962 R. Kaprelyan's crew lifted a 20,117 kg load to an altitude of 2,738 meters. In September 1961 the crew commanded by N. Leshin developed the speed of 320 km/hr at 15-25 km altitude. It was an absolute world record in speed for a rotary-wing aircraft. Such a high speed had been believed to be unachievable for that type of aircraft. The meters. Mil Design Office was awarded with the I. Sikorsky World Prize.

The Mi-6 helicopter took part in air parades and was demonstrated at international aviation shows repeatedly.

In 1968 test pilots V. Koloshenko and Y. Garnayev flew the helicopter to Western European cities.

The Mi-6 is widely used in the Air Force as well as in the national economy of our country. It is also operational in a number of states abroad.

The helicopter exhibited in the Museum was built in the late 1960s.

Till June 1963 it was at the factory. It flew 70 hours 14 minutes. It performed 354 aircraft and 43 helicopter landings. Total flight time is 96 hours 14 minutes. The helicopter was delivered to the Zhukovsky Air Force Engineering Academy training base on May 26, 1965, and it became an exhibit in 1978.

# Mi-10 Helicopter

In the late 1950s the Mil Design Office developed the Mi-10 helicopter designed for airlifting large-size cargoes. The power plant is comprised of two D-25V gas turbine 5,500 HP engines. The helicopter is provided with a special platform.

The extended undercarriage legs enable the helicopter to taxi over a cargo up to 3.5 m high. Special hydraulic lifts enable it to pick up loads from the ground.

The first helicopter flight was performed by test-pilot R. Kaprelyan on October 15, 1960. The Mi-10 tests were going on until the end of 1963, having demonstrated the machine's capability to airlift loads 20 m long, 5 m wide, and 3.1 m high, and weighing up to 12,000 kg over a distance of 250 km (with additional fuel tanks - up to 630 km.), and a 15,000 kg load at a short distance of not more than 25 km.

The Mi-10 take off weight (in vertical take-off) is 43,450 kg. Cruising speed with a cargo on the platform - 180 km/hr.

In 1965 the Mi-10 set a number of world records: 25,100 kg cargo was lifted to 2,480 meters altitude; 5,000 kg cargo was lifted to 7,151 m altitude.

The Mi-10K with shortened undercarriage legs and an additional carbine under the nose made in 1966 was designed for construction and erectional jobs.

The helicopter was provided with a triple control system. For performing erectional functions, one of the pilots takes a seat in the lower cabin, faces the cargo to be loaded, and controls the operation.

In 1967 the helicopter was demonstrated at the aviation show in Paris. The helicopter revealed its abilities by dismantling one of the Paris bridge crane eleven ton frameworks.

The Mi-10 helicopter exhibited in the Museum was produced in November 1968. The helicopter flight readiness tests were performed on November 11, 1968.

The helicopter was in service from 1968 till 1974. Its flight time is 92 hours. It was flown for the last time by pilot Pelevin, and landed on the Museum airfield on December 25, 1974.

# Mi-8 Helicopter

In May 1960 the Mil Experimental Design Office began developing the Mi-8 helicopter powered by two S. Izotov TV2-117 gas turbine engines with 1,700 HP each. Five-blade rotor is 21.3 m in diameter. Maximum and normal take-off weight - 12,000 kg and 11,000 kg respectively. The maximum eight of the load carried internally is 3,000 kg. It can carry up to 3,000 kg externally. Loading and unloading is performed along the loading ramp in the rear cargo compartment.

The cargo compartment ensured airlifting of 20 soldiers (max 24 soldiers) with individual weapons and equipment to a distance of 450 km. The ambulance version cargo compartment accommodated 12 lying wounded or sick and nurses.

The combat version carried 4 rocket pods internally.

The passenger version compartment accommodates 28 passengers.

The helicopter was provided with an improved autopilot, and the main rotor rotation automat responsible for the rotation constancy as well as for synchronization of the engine speed. The flying-navigation systems had day/night adverse weather capability. The Mi-8's first flight was made by test-pilot N. Leshin on September 17, 1962.

During the tests the helicopter revealed the following characteristics: maximum speed - 250 km/h; maximum altitude - 4,500

meters; range of flight with full fuel tanks and maximum load - 400-450 km; three-man crew: and two pilots and a mechanic.

The helicopter entered serial production in late 1962. It is widely used in various branches of the national economy, as well as abroad. The Mi-8 set 8 world records.

The Mi-8 (No. 0606) exhibited in the Museum was produced in July, 1966.

On September 30, 1966, Pilot Farskin ferried it to an air unit, where it flew 737 hours. As soon as the machine flying time came to an end it was delivered to a repair shop. After the overhaul the helicopter flying time counted 500 hours. On September 30, 1971, it was taken into service by an air unit in which it flew 220 hours.

On April 30, 1971, the helicopter was taken by another air unit. It flew 12 hours there, and on May 24 of the same year it was given to the Zhukovsky Air Force Engineering Academy training base. On August 16, 1978, it was delivered to the Museum.

The total flight time is 983 hours, though after the last repair it is 232 hours.

# Mi-2 Helicopter

The experience of installing gas-turbine engines on helicopters proved their advantage over the original piston engines: light weight, relative cheapness of fuel, and the ability to develop higher performance.

For these reasons the Mi-1 was replaced by the three-blade rotor (14.5 m in diameter) light Mi-2 powered by two GTD-350 Izotov 400 HP turboshaft engines installed in the upper fuselage. These engines enabled the enlargement of the cabin, accommodating 8 passengers or up to a 800 kg load. The helicopter's take-off weight with such a load is normally 3,700 kg.

The Mi-2 was first flown by test-pilot Alfyorov on September 22, 1961.

During the tests the helicopter revealed the following performance: maximum speed - 210 km/hr, static ceiling - 1,450 meters, dynamic ceiling - 4,000 meters, range of flight - 580 km. Crew - one man.

The Mi-2 versions entered serial production. It was widely used in national economic spheres. In the Land Forces aviation it performs chemical reconnaissance and artillery fire adjustment missions.

The Mi-2 (No. 510309037) exhibited in the Museum is an ambulance helicopter. It was produced in March 1967. Until June 28 of the same year it was tested at the factory. Later, it was oper- ated in an air unit, and then at the Moscow Aviation Plant test base. The last flight was made by pilot Volin on August 19, 1975. The total flight time is 220 hours 16 min. It was delivered to the Museum on November 20, 1975.

# Mi-12 Helicopter

In 1965 the Mil Design Office was assigned the mission of developing a helicopter for airlifting no less than 30,000 kg of load.

The one-blade scheme traditional for the Design Office couldn't meet the requirements of creating a reductor capable of transmitting over 20,000 HP to the rotor, which was necessary for such a helicopter. That is why the Mi-12 became the first and the only two-rotor transverse scheme helicopter in the Mil Design Office. Both rotors, 35 m in diameter each, were installed on the wing tips. Additionally, there are two D-25 VF Solovyov gas-turbine engines, 6,500 HP each.

The synchronization of both rotors' rotation was ensured by the transmission shaft connecting both reducers. The shaft is also used for transmitting power from one reducer to another when the left and the right engine groups possess different powers due to one or even two engine failures.

Unlike an aircraft, the Mi-12 wing is tapered towards the fuselage. This produced a reduction of the air blast around the wing coming from the rotors, since the most powerful blast arises near the fuselage. The helicopter is provided with a tail unit used at a

translational velocity. The fin is provided with an elevator which changes its position as soon as the rotors' pitch does.

The ground-based tests begun in June 1967 revealed the helicopter had resonant vibrations which were soon eliminated. The flying tests were performed by test pilot V. Koloshenko.

The helicopter delivered the following performance: maximum speed - 250 km/hr; service ceiling - 3,500 meters; maximum take-off weight - 105,000 kg.

The crew consists of six men: two pilots, a navigator, a radio operator, an electrician, and an engineer. The navigator and the radio operator are in the upper part of the double-level cockpit.

In February 1969 during the flight tests the Mi-12 lifted a record load of 31,030 kg to 2,951 m altitude.

The Mi-12 didn't enter serial production. Only three were built. In 1975 one of them was flown to the Museum.

# Mi-24A Combat Helicopter

In 1972 the Mil Experimental Design Office created the Mi-24A anti-tank armor support helicopter. This was the further development of the Mi-24 combat helicopter's family.

The helicopter is powered by two TB-3-117A gas-turbine engines with 2,200 HP each. The main rotor is 17.3 m in diameter. Normal take-off weight - 10,500 kg.

During the test the helicopter demonstrated the following characteristics: maximum speed - 340 km/hr, dynamic ceiling - 5,100 meters. The 1,710 kg fuel endurance ensures 745 km range of flight and one hour twenty minutes' flight time.

Armament: four anti-tank guided missiles; four UB-32 pods with 32 unguided rockets each; one 12.7 mm A-12 machine gun (900 shells). The pods may be replaced by four FAB-100 bombs or two OFAB-500 bombs.

The Mi-24 modifications entered serial production. They were used in combat operations and proved efficient.

The very first Mi-24 flight was performed by G. Alfyorov on September 19, 1969. The Mi-24A (No. 2201201) exhibited in the Museum was produced in August 1972. Since February 16, 1973, the helicopter has been in service with air units.

On June 27, 1984, pilot Pozdnyakov landed the helicopter on the Monino airfield and delivered it to the Museum.

The helicopter's total flight time is 780 hours, though after the last overhaul it counts 252 hours. It performed 935 landings.

# Ka-15 and Ka-18 Helicopters

While the single-rotor type helicopter is traditionally typical of the Mil Experimental Design Office, the Kamov deals with the coaxial-rotor helicopters. Kamov explained his conception in the following way, "Almost complete symmetry of the main rotor system, the shortest inertia moment, no losses of power for the tail rotor rotation, small sizes of the helicopter."

In 1947 the Ka-8 coaxial-rotor type helicopter powered by a motorcycle engine was designed. The Ka-10 helicopter of a similar type, powered by the AI-4V air-cooled engine, was designed in 1949.

The Ka-15 helicopter was developed in 1954. It was a coaxial-rotor type helicopter, the main rotors being 9.95 m in diameter. The

helicopter was powered by the Ivachenko AI-14V aircooled engine. The cockpit accommodated a pilot and two passengers.

The Ka-15's first flight was performed by test-pilot D. Yefremov on April 13, 1953. The helicopter underwent production and official flight tests successfully. It flew at 150 km/hr at 3,000 m altitude, with a take-off weight - 1,480 kg. The Ka-15 set a record when it developed a speed of 170 km/hr in an enclosed circuit route of 500 km. The helicopter designed for the national economy employment was put into serial production.

The Ka-15 No.1 helicopter exhibited in the Museum was built in March 1953. The ground tests began on March 28, 1953. The first flight was performed on April 13, 1953.

In April 1956 the helicopter was given to an organization. It was flown there up to 1959. The last flight was made by pilot Yarkin on April 6, 1959. The helicopter flight time is 1,474 hours. It was delivered to the Museum from the helicopter factory on July 7, 1977.

In 1956 the Design Office developed the Ka-18 helicopter, which was equipped with the same engine and rotors as the Ka-15. The helicopter structure enabled the accommodation of up to 3 passengers in the cabin. Ka-15 passenger, ambulance, and agricultural versions were developed.

Performance: maximum speed - 145 km/hr; dynamic ceiling - up to 3,000 meters; flying mass - 1,460 kg.

In 1958 the helicopter mass was demonstrated at an air show in Brussels and was awarded with "The Gold Medal." It was dismantled and delivered to the Museum on September 2, 1975.

# Ka-25 Antisubmarine Helicopter

The Ka-25 carrier-based antisubmarine coaxial twin-rotor helicopter was developed by the Kamov Design Office in 1957. The rotor diameter is 15.7 m. The rotor blades can fold when on the ground. The power plant includes two GTD-3 S. Izotov engines with 900 HP each.

The helicopter displayed the following characteristics: speed - 220 km/hr; ceiling - 3,500 meters; range of flight - 650 km; and take-off weight - 7,300 kg.

The helicopter is fitted with electronic and radar equipment, enabling it to communicate with ships in combat operations as well as to perform search and rescue.

The Ka-25 helicopter (No. 5910203) exhibited in the Museum was built in December 1966. It underwent flight tests at the factory. It remained there until August 9, 1979. Pilot Isayev flew it for the last time on August 29, 1975.

The total flight time of the helicopter is 314 hours.

The helicopter was delivered to the Museum from the Kamov Design Office on January 5, 1981.

# Ka-26 Helicopter

In 1965 the Kamov Experimental Design Office developed and built the Ka-26 coaxial rotor type helicopter. The airscrew diameter is 13 meters. The propulsion system is comprised of two M-14V-24 aircooled piston engines with 325 HP thrust each.

The Ka-26 versatile helicopter is designed for airlifting passengers and relatively light loads as well as for agricultural jobs.

Structurally, the helicopter may be regarded as a "flying undercarriage" on which one can install either a six-passenger cabin or a large capacity bin for fertilizers and other transportable substances, or a 900 kg cargo platform.

The Ka-26 prototype was first flown by test-pilot V. Gromov on August 18, 1965. It underwent tests successfully and displayed the following characteristics: speed - 170 km/hr; static ceiling - 640 meters; dynamic ceiling - 2,100 meters; and take-off mass - 3,250 kg. Fuel endurance allows it to cover a distance of 520 km.

The Ka-26 entered serial production and is widely used in various branches of the national economy.

It was demonstrated at international shows in 22 countries. It was awarded two Gold Medals in Moscow and Plovdivm and the Grand-Prix Silver Cup in Budapest, as well as with a great number of diplomas. The Ka-26 helicopter is widely used in Japan, Sweden, FRG, and other countries.

The helicopter exhibited was delivered to the Air Force Museum from the Kamov Factory on June 4, 1986.

# Yak-24 Assault Helicopter

In late 1951 the Yakovlev Experimental Design Office (P.D Samsonov, N.K. Skrzhinsky, I. E. Erlikh, and some others) began developing a heavy troop-carrying transport helicopter designated the Yak-24.

The Yak-24 has two four-bladed tandem rotors 21 m in diameter. The helicopter is also equipped with ASH-82V Shvetsov piston engines, with 1,430 HP normal rated power and 1,700 HP take-off power each.

To reduce the influence of the main rotor on the tail of the helicopter, one was installed at a considerably higher level than the other.

The helicopter transmission included two main and two intermediate reducers, and transmission shafts connecting the engines and the main reducers, thereby ensuring synchronous oncoming rotation of the rotors.

The fuselage had a special form covered with a linen skin.

The front fuselage accommodated a three-seat cockpit (2 pilots and a radio operator). There was a provision for a machine-gun in the cockpit.

The Yak-24 was believed to be used in three versions: troop-carrier, transport, and ambulance.

The troop-carrying version accommodated up to 40 soldiers.

The ambulance version cargo cabin accommodated 18 stretchers, a table, and a seat for a nurse.

The transport version cargo cabin enabled airlifting 4,000 kg load internally and 3,500 kg externally. In the rear cargo cabin there

is a hinged loading ramp. In the middle of the deck there is a loading hatch used in hovering. Inside the cargo cabin an electric hoist with 200 kg lifting power can be installed.

On July 3, 1952, the helicopter crew comprised of test-pilots S. Brovtsev and E. Milyutichev flew the Yak-24 for the first time.

In subsequent flights the test-pilots experienced serious vibrations, which threatened the helicopter's structural integrity. The real reason for that phenomenon turned out to be the unfavorable combination of fuselage deformations and poor arrangement of the control operating rods.

As a result, a number of structural changes were implemented, and the vibrations were reduced. By the end of 1954 the production flight tests had ended. Then the helicopter underwent official tests that were completed in April 1955. The Yak-24 was highly appreciated and put into serial production.

During the tests the helicopter displayed the following characteristics: maximum speed - 175 km/hr, static ceiling - 2,000 meters, zoom altitude - 4,200 meters, range of flight - 265 km, take-off weight - 14,270 kg. The tandem system, namely the low sensitivity to the centering change, proved to be advantageous.

The Yak-24 surpassed all Soviet and foreign helicopters of those years in engine power and payload capacity.

In December 1955 two world records were set when pilot E. Milyutichev flew the Yak-24 with a 4,000 kg load and reached 2,904 m altitude, while pilot Tenyakov lifted a 2,000 kg load to 5,082 m altitude.

The Yak-24 was the first helicopter in our country to be used as a flying crane. In June 1959 the Yak-24 was used in the restoration of the Yekaterininsky Palace in the town of Pushkin. In July of the same year it was used in constructing the Serpukhov-Leningrad gas pipe.

The Yak-24 (No. 03310) exhibited in the Museum was produced in 1956. It flew 78 hours. The last flight was made in December 1958. It was delivered to the Museum on March 14, 1962.

# 7

## Lightweight Aircraft and Trainers

## The "Burevestnik" (Stormy Petrel) Light Aeroplane

In 1922-1923 Vyacheslav Nevdachin—one of the first Russian pilots and air engineers—together with the members of the Air Fleet Amateurs Society, built a glider and four light airplanes designated "Burevestnik" (Stormy Petrel).

In 1923 the P-5 Burevestnik glider participated in the first All-Russia glider competition after which Nevdachin strengthened its structure and installed the Harley-Davidson motorcycle 7-9 HP engine. The engine was slightly rebuilt, but it was not powerful enough.

The light airplanes were improved, the engine power increased, and the sizes reduced. The performance improved, and the C-4 Burevestnik light aeroplane powered by the Blackburn-Tomtit engine operated perfectly well.

The turn time was 14 seconds, take-off ground roll and landing roll distance - 30 meters, speed - up to 140 km/h. It performed the Moscow-Odessa flight. The Burevestnik was able to perform aerobatics.

The airplane was an all-wooden aircraft with linen skin. Wing span - 9.78 meters. length - 5.7 meters, maximum speed - 140 km/hr.

On July 1927 pilot Zhukov set a 5,500 m altitude record in it.

The "Burevestnik" exhibited in the Museum was delivered from the Frunze Central House of Aviation and Cosmonautics on October 15, 1963. It was in bad repair and without any certificates. It was assembled and completely restored by the museum brigade of aircraft mechanics headed by A. Mansurov.

# UT-2 Trainer

By the mid-1930s the U-2 Soviet primary trainer, which was to stay in service during both peace and wartime, had become obsolete. The adoption of new high-speed combat monoplanes created a significant gap between a trainer and a combat aircraft. The Air Force required a high-speed trainer with the scheme corresponding to that of a combat monoplane. The UT-I single-seater and the UT-2 two-seater were developed by the Yakovlev Design Office in 1937.

The UT-2 was the AIR-10 modification. It was powered by the M-II 100 HP engine. It developed a speed of 205 km/hr. Flight endurance was 7 hours, ceiling - 3,500 meters, flight weight - 940 kg, length - 7m, wing span - 10.2 meters. It was first flown by Yu. Piontkovsky. The UT-2 was easy and cheap for production. It was put into serial production in September 1937. A total of 7,150 UT-2s were produced. The UT-2 was the primary trainer from 1938 until 1948. Hundreds of thousands of military pilots were trained on it. A great number of UT-2s were used in flying clubs up until 1950. The UT-2 exhibited in the Museum was delivered there from the Yakovlev Experimental Design Office on July 5, 1978.

# Yak-12R Versatile Light Aeroplane

The aeroplane is the Yak-12 modification made in 1946.

The Yak-12R is equipped with an AI-14R 260 HP engine comprised of the VISh-530 L-11 propeller and R-2 propeller controller.

Maximum flying weight - 1,221 kg, payload capacity - 1,305 kg, maximum speed - 184 km/h, landing speed - 62 km/h, take-off run - 60 meters, landing run - 86 meters. The aircraft is also equipped with the RSI-6 radio station. The aircraft can operate at night and under any weather conditions. The Yak-12R was adopted for service as a liaison aircraft. Later it was used for glider towing and parachute jumps at flying clubs.

The Yak-12R exhibited in the Museum was produced on October 1, 1955. The aircraft flew 173 hours and performed 80 landings. The aircraft and its engine were written off on December 28, 1961. The aircraft was delivered to the museum by the military department of the Moscow Physiotechnical Institute on January 11, 1962.

# Yak-11 Fighter-Trainer

The Yak-11 two-seat trainer was designed by the Yakovlev Design Office in 1946. The aircraft was to possess similar handling characteristics to those of propeller-driven aircraft of the same design, but was specified to be less expensive in maintenance.

The Yak-11 is a low-wing cantilever monoplane with a retractable undercarriage.

Length - 8.503 meters, wing span - 9.4 meters, height - 3.283 meters. Empty weight - 1,811 kg, payload with bombs 669 kg. Maximum speed at 2,250 m altitude - 475 km/hr. Ceiling - 7,500 meters. Time to climb - 36.8 min.

It entered serial production in the USSR and Czechoslovakia. A total of 3,860 aircraft were produced. The Yak-11 is powered by the ASh-21 seven-cylinder radial direct injection 700 HP engine. The engine was equipped with a direct injection set - the NV-21 pump. The metal two-blade variable-pitch propeller was provided with the R-FE constant rotation regulator.

The aircraft is equipped with a gunsight, the PAU-22 aerial combat camera (enabling it to perform firing practice against air and ground targets), and two beam bomb racks able to carry one 50 kg bomb each.

The Yak-11 aircraft is equipped with the RSI aircraft simple high-frequency radio/telephone station, radio semi-compass, and the SPU-2M bis intercommunication set.

In the period of 1949-1951 the aircraft underwent a number of structural changes. It was provided with a new sight. The PAU-22 installed in the wing was replaced by the C-13 aerial combat camera on the windscreen. The AGP-2 gyro horizon was replaced by the AGK-47B electrically-operated combined gyro horizon.

The Yak-11 was widely used in air units and flying schools. In 1956 the Yak-11U nose wheel modification was implemented.

The Yak-11 exhibited in the Museum was made on February 5, 1954. The aircraft flew 709 hours and performed 2,978 landings. It was delivered to the Museum on May 27, 1965.

# Yak-18 Sport Trainer

In the postwar period there was a tendency in the Soviet Air Force to design sport trainers similar to combat planes. In 1946 the UT-2 was replaced by the Yak-18 sport trainer possessing a similar lay-out and provided with retractable landing gear, high-lift devices, variable pitch propeller, and enclosed cockpit.

In the same year the aircraft entered serial production and soon became the primary trainer at flying schools and clubs. A total of 7,000 Yak-18 aircraft of different modifications were produced.

They were widely used in our country as well as abroad in the GDR, Austria, Poland, etc.

The airplane has an all-metal construction with the tail unit and outboard wing panel having a linen skin. The cockpit is provided with a plexiglass canopy. The aircraft is equipped with the engine double control system, and all the accessory equipment facilities except for the undercarriage crane installed in the forward cockpit. It has the M-11 FR 160-HP engine, and a VISh-327 EV-249 two-blade metal variable pitch propeller.

Maximum speed - 250 km/h. Ceiling - 4,000 meters. Range of flight - 1,010 km. Take-off distance - 290 meters. Landing distance (with flaps and brakes) - 250 meters. The aircraft can operate day and night under adverse weather conditions. The radio equipment enables two-way communication with the ground. It is comprised of a RSI-6MP receiver, RSI-6K transmitter, and a SPU-2M bis intercommunications set.

The Yak-18 exhibited in the Museum was produced on March 17, 1954. It was delivered to the Museum from the Moscow aircraft repairing base on November 27, 1958. The aircraft time is 2,127 hours. It underwent 5 serious overhauls.

# Yak-18U Trainer

It was one of the Yak-18 modifications used for primary flight training. It was designed in 1954. It is an all-metal two-seat low-wing monoplane since its undercarriage is retractable.

The aircraft is powered by the Shevtsov M-11FR 160 HP engine, and possesses the V-501 D-81 variable pitch propeller 2.3 m in diameter.

The aircraft is equipped with the RSI-6K HF radiostation, RPKO-10M radiosemicompass, and SPU intercommunications set. It possessed practically the same flying and handling characteristics as those of the Yak-18.

The Yak-18U exhibited in the Museum was produced on December 1, 1955. It flew 1,059 hours from the Moscow Physiotechnical Institute military department on January 11, 1962.

# Yak-18PM Acrobatic Aircraft

In the late 1940s and early 1950s Yakovlev designed a number of sport aircraft, the Yak-18P single-seat acrobatic plane among them.

The aircraft presented itself in a good light. It was designed for advanced acrobatic flying. Having taken into consideration its good performance and sharp needs of flying clubs for sport aircraft, DOSAAF ordered 100 aircraft. The successful participation of the Yak-18P Soviet pilots on the Second and Third World Championships in Hungary and Italy (1962 and 1964) proved the high reputation of the aircraft as an aerobatic machine.

The next modification of the aircraft was the Yak-18PM produced in 1965. At the Fourth World Aerobatics Championships held in Moscow in 1966, both the male and female pilots of the Yak-18PM won all the gold, silver, and bronze medals. They were pre-sented with the Nesterov Cup for team victory, the Yak-18PM being acknowledged as the best sport aircraft.

The Yak-18PM is a sport aerobatic plane. It's powered by a Ivachenko AI-14RFP 285 HP engine. Its take-off weight is 1,100 kg. Maximum speed is 320 km/h with a landing speed of 115 km/h. Service ceiling of the Yak-18PM is 6,700 m and range of flight is 400 km. Take-off and landing distances are 142 m and 133 m respectively.

The aircraft was provided with the R-860 radio station.

The Yak-18PM exhibited in the Museum is the World Aerobatics Championship participant, held in Spain in 1966. Such pilots as V. Martemyanov, I. Yegorov, G. Korchuganova, and S, Savitskaya flying this type of aircraft won the World Advanced Aerobatics Championships. The aircraft was delivered to the Museum from the V. Chkalov Central Flying Club.

# Yak-18T Versatile Aircraft

As a result of successful employment of the Yak-18 trainer family Yakovlev designed a versatile Yak-18T aircraft in 1967. The aircraft was designed to train pilots and parachute jumpers as well as for airlifting either 3 passengers or a 400 kg load on local air routes. The cockpit arrangement enables it to be used as an ambulance. The two-seat version is capable of performing all the aerobatics.

The Yak-18T is an all-metal low-wing cantilever monoplane with a nose wheel retractable gear. Flying weight is 1,620 kg.

The aircraft is powered by the AI-14 300 HP piston engine. The two-blade propeller possesses a propeller rpm governor.

Maximum speed is 300 km/hr. Service ceiling is 4,000 meters. Take-off distance is 300 meters, landing distance is 290 meters, range of flight is 1,000 km.

The aircraft was delivered to the Museum from the Yakovlev Design Office in non-flying condition on March 1, 1974.

# Yak-30 Sport Aeroplane

By the early 1960s, it became necessary to design not only combat and transport jet aircraft, but sporting ones as well. The Yakovlev Design Office developed two versions of such an aircraft: the Yak-30 two-seat and the Yak-32 single-seat aircraft, powered by RU-19 engines with 30 kg thrust each. Flying these machines, sporting pilots G. Korchuganova, R, Shakhina, V. Smirnov, and V. Mukhin managed to set a number of world records.

The Yak-30 take-off weight is 2,250 kg. (The Yak-32 - 1,930kg).

Maximum speed is 660 km/hr, landing speed is 140 km/hr, service ceiling 14,000 meters, range of flight - 965 km.

The Yak-30 exhibited in the Museum was produced by the Yakovlev Design Office factory. It was delivered to the Museum in a non-flying condition, though equipped with complete standard equipment, on April 1, 1975.

# Yak-50 Sport Aeroplane

The Yak-50 single-seat all-metal sporting aeroplane is the further development of the Yakovlev aerobatic aircraft family.

It is powered by the M-14P nine-cylinder 360 HP engine and provided with a two-blade propeller possessing a propeller rpm governor.

The Yak-50 take-off weight is 900 kg, maximum speed - 300 km/hr, ceiling - 5,500 meters, G-load range is 0-6, and fuel capacity is 551. There is provision for an external fuel tank allowing the aircraft to cover a distance of 500 km.

The aircraft is equipped with the HF "Zyablic" radiostation.

The Yak-50 entered service in 1974. Flying the Yak-50 aeroplane the Soviet team won the Mesterov World Cup and 23 medals out of 30 at the 8th World Championship. Pilots V. Letsko and L. Leonov became world champions. They repeated their success at the European Championship in France in 1977.

The Yak-50 airplane exhibited in the Museum was made at the Yakovlev Design Office factory. It was delivered to the Museum from the factory in flying condition on April 15, 1975.

# Yak-52/B Primary Trainer

In 1974 Yakovlev designed the Yak-52 two-seat primary trainer. Its further modification became the Yak-52B, enabling it to conduct piloting practice as well as combat employment training. The aircraft possesses a sight and two bomb racks with UB-32 bombs, enabling it to perform mock firing against moving and stationary ground and low-level air targets.

Take-off weight is 1,220 kg. The aircraft is powered by the M-14P 360 HP engine. Maximum speed is 280 km/h; flight endurance is 2 hours 50 minutes. Take-off distance is 200 meters; landing distance - 260 meters.

The aircraft was in serial production until 1981.

The aircraft exhibited in the Museum began flying on April 20, 1978, and ended its career on October 28, 1983. The total flight time is 125 hours. It performed 202 landings. It was delivered to the Museum from the Yakovlev Design Office factory on July 6, 1987.

# L-29 Dolphin Trainer

In Autumn of 1961, competitive flight tests of three primary trainers represented by 3 countries took place. The countries competing were the USSR, Czechoslovakia, and Poland. The trainers tested: Yak-30, L-29 Dolphin, and TS-11 Spark. The tests were performed on the Monino airfield.

The task was to choose an aircraft that would meet primary trainer requirements, which included ease of control, reliability, and the capability of operating from grass strips. The competitive characteristics and flying tests proved that the Czechoslovakian L-29 Dolphin met those requirements best of all.

The L-29 became the first jet primary trainer in the USSR. The first shipments of the aircraft were delivered to the Soviet flying schools in August 1963. Much was to be done for conversion training of pilot-instructors and engineering technical staff. In April 1954 the aircraft was first flown by cadets.

The L-29 is an all-metal two-seat monoplane. Its length is 10.8 meters. Wing span is 10.3 meters. Take-off weight is 3,344 kg. The plane is powered by M-700BC-18 900 kg static thrust engine.

The total fuel capacity is 1,030 l. Maximum speed is 630 km/hr. Maximum operational overload is 8. Take-off run on a surfaced strip is 600 meters, landing run is 500 meters. Service ceiling is 10.9 km.

The armament includes FkP-2-2 gun-camera, ASP-3 NM sight, and a bombardment and missile-launching equipment. Bomb load is 2x100 kg. The seats are ejectable. The aircraft is provided with a HF RTL-11, ARK-9 radiostation.

The aircraft exhibited in the Museum was delivered from the Zhukovsky Air Force Engineering Academy training base on December 14, 1973.

## "Stork-2" Superlight Aeroplane

In the late 1960s a superlight airplane building campaign took place. This aircraft had both ancestors (V. Nevdachin and I. Tolstykh light airplanes) and successors (MAI-890, etc). Such planes were powered by improved low-power motorcycle engines.

The Stork-2 airplane was built by engineer V. Vyugov with a group of schoolboys.

The Stork-2 is a single seat high-wing monoplane designed for air-field flights. The plane is powered by the Ural M-62 motorcycle two-cylinder aircooled 28 HP engine. The propeller diameter is 1.25 meters.

The plane weight is 265 kg. Payload is 85 kg. Wing span is 8.8 meters. Length is 5.7 meters. Aerodynamic efficiency is 11. Maximum speed is 100 km/hr, with landing speed - 60 km/hr.

The plane performed 337 flights. Its flying time is 125 hours. The plane climbed up to an altitude of 410 meters.

In 1980 the plane was displayed at the Youth Scientific and Technical Exhibition from which it was delivered to the Museum.

# 8

## Unpowered Flying Vehicles

## A. Tupolev Glider (Mock-up)

The glider was constructed by A. Tupolev during his student years in 1912 and the future designer performed his first flight over the Yauza river.

The mock-up is made of wood and has metal braces.

The length is 4m, wing span - 5.5 meters, weight - 45 kg. The undercarriage - metal skis.

The mock-up in the AF Museum was made for the day of A. Tupolev's 100th anniversary and delivered to the Museum on July 3, 1990.

## P. Ivansen "Joseph Unshlikht" Glider

"Joseph Unshlikht" glider ("Sea-gull") was designed by engineer P. Ivansen in 1934-35 and constructed at the Research Institute of Civil Aviation in 1935. It is a single-seat, record-breaking open-type glider designed for soaring flights.

The wing span is 30.4 meters. Aerodynamic efficiency - 29. Flying weight - 260 kg. It is a cantilever high gull-wing monoplane of wooden structure.

The glider was received by the AF Museum from the Central House of Aviation and Cosmonautics named after meters. Frunze on August 22, 1961.

# O. Antonov A-11 Glider

The A-11 glider is a single-seat high-speed training and record-breaking soaring glider. It was constructed in 1956 by the O. Antonov Experimental Design Office.

It was series produced until 1960. The first flight was performed by C. Anokhin on the 12th of July, 1958. The glider is an all-metal (for the first time in the USSR), cantilever high-wing monoplane, with a V-tail and retractable undercarriage.

Main glider characteristics: wing-span - 16.5 meters, length - 6 meters; wing area - 12.15 square meters; maximum L/D ratio - 28; maximum speed - 250 kph; flight mass - 410 kg; structural mass - 320 kg.

The glider set some all-Union records, including flight range.

# B. Sheremetev "Kashuk" Glider

The "Kashuk" glider was designed by B. Sheremetev and constructed in 1952 by engineer A. Monopkov. It was named "Kashuk" after Oleg Koshevoi - one of the participants in the underground resistance group in the Ukraine during the Great Patriotic War.

The "Kashuk" glider was the world's first flapping wing glider tested in flight. The incoming airflow raises the movable wing, com-pressing the air in a special compressor, and then the wing resets with the help of compressed air. The pilot can have the glider's wing fixed in the initial position with the help of a special device.

Main glider's dimensions: wing span - 17m; wing area - 15.1 square meters; maximum L/D ratio - 30; take-off mass - 400 kg; strength factor - 8.

# O. Antonov A-13M Powered Glider

The A-13M powered glider was a light single-seat jet glider which originated from the A-13 glider in the O. Antonov Experimental Design Office in 1959 and was designed for performing aerobatics and setting speed records. It is an all-metal monocoque structure.

It has a TS-31 engine with a thrust of 50 kg, which makes it possible to take off and gain speed without a towing vehicle. The first flight was performed by S. Anokhin on February 11, 1960.

Wing span - 12.1 meters, wing area - 10.44 square meters, maximum lift-to-drag ratio - 23.5, length - 6 meters, speed with max quality - 125 kph, take-off mass - 499 kg. Maximum speed - 300 kph. Design G-load - 9.5.

# O. Antonov A-15 Glider

The A-15 was a single-seat record-breaking soaring glider designed by the collective of O. Antonov Experimental Design Office. It is an all-metal cantilever high-wing monoplane with V-tail.

Designed in 1960, it was series-produced. The first flight was performed by S. Anokhin on March 26, 1960. It has powerful high-lift devices (extension slotted flaps and dropped ailerons providing for the minimum speed with short radius turn flights). The glider has a full package of navigational equipment which makes it capable of flying under instrument weather conditions.

Main glider dimensions and performance: wing span - 18 meters; wing area - 12.3 square meters; maximum L/D ratio - 39; length - 7.2 meters,; max speed - 250 kph; minimum speed - 65 kph; take-off mass - 420 kg.

The A-15 glider has set four world records and 26 all-Union records. It was one of the best soaring gliders constructed in our country. The A-15 glider exhibited at the AF Museum was received from O. Antonov Experimental Design Office and delivered dismantled on February 4, 1974.

# KAI-19M Glider

The KAI-19M is a single-seat, record-breaking open type soaring glider designed for soaring flights under simple weather conditions and in air turbulence. Due to flight-control and navigation equipment it can perform soaring flights in clouds. The glider was received from the Sport Aviation Experimental Design Office on January 13, 1975.

It is a cantilever high-wing monoplane of all-metal structure with a T-tail and single-wheel landing gear.

The cockpit is designed to accommodate the reclining position of the pilot. The glider was built by Sport Aviation Experimental Design Office (Kazan) in 1957. Tests were conducted until 1971.

Wing span - 20 meters; wing area - 14 sq. meters; L/D ratio - 41; length - 8m; maximum flight speed - 250 kph; minimum speed - 55 kph; take-off mass - 460 kg.

# "Letuva" LAK-9M Glider

The "Letuva" LAK-9M glider was designed and built at Pryanaisik factory of sport aviation by the collective of the Karvyalis Balis Design Office in 1977.

The glider is constructed of plastic and has the following characteristics: wing span - 20.2 meters; wing area - 15 sq meters; fuselage length - 7.27 meters; maximum L/D ratio - 48; take-off weight - 450-630kg; optimum glide speed - 103 kph.

The "Letuva" LAK-9M glider (No. 408) was delivered in full package to the AF Museum by the sport aviation factory on April 18, 1983.

# 9

## Discoplanes/Deltaplanes

## M. Sukhanov Discoplanes

The idea of a discoplane, an aircraft with a round-plan-view wing was suggested more than once by the designers of different countries. However, the experience of their construction and flight tests was highly limited.

Discoplanes I and II—experimental gliders with disk wings—are designed for investigations of the given arrangement influence on gliding descent spans at a low altitude envelope. They were designed by meters. Sukhanov. Number I—of great size—was built in 1950 in Novosibirsk, while Number II was built in 1960 at the factory named after S. Lavochkin.

The wing diameter of the first discoplane was 5 meters, the second, 3.45 m, and the wing areas were 20 square m and 9 square m respectively. Maximum L/D ratio - 8. Flight speed - 60-22 kph. Descent velocity - 5m/sec. Aerodynamic control devices-elevons and turn surfaces.

The first discoplane was tested by the designer himself. The remaining flight tests were conducted by the master of sport gliding, V. Ivanov, in August of 1957. Nine flights were performed with towing by Yak-12 to an altitude of 2,000-2,500 meters. The glider demonstrated fine flight performance. Absolute anti-spinning is the most valuable performance attribute of the discoplane. The "air cushion" effect ensures safety of landing with 45 degree wing angles of attack.

Both positive and negative (small L/D ratio) qualities of the round-plan-view wing were tested.

In 1979 meters. Sukhanov continued to develop the idea of the discoplane on the theory that it could be used for the recovery of spaceships in orbit.

# O. Antonov UT-03 "Slavutich" Deltaplane

The UT-03 "Slavutich" Deltaplane was worked out by O. Antonov Design Office, and was designed for primary training at flying clubs and participating in competitions and record-breaking flights.

In the process of its creation, configuration and strength tests were conducted in the wind tunnel and in flight. The UT-03 "Slavutich" deltaplane demonstrated good performance and was recommended by the DOSAAF Central Committee for flights as the primary trainer at DOSAAF clubs.

"Slavutich" mass - 25 kg. Wing area - 17.5 square meters; length - 4.30 meters. Wing span - 8.90 meters. Design G-load - 4.5. Minimum descent velocity - 1.3 m/sec. Speed - 7-20 m/sec.

The exhibited deltaplane was received from the O. Antonov Experimental Design Office on August 20, 1984.

# "Albatross-20" Deltaplane

The "Albatross-20" Deltaplane was designed and built by the students of the Bauman Moscow Higher Technical School "Albatross" club in 1979. They were S. Stoiko and V. Gulaikin.

It is the "Mirage" vehicle prototype which was built by the members of the "Semal" club in Alma-Ata and designed for sporting and record-breaking flights over mountains.

The deltaplane mass is 23 kg. It is made of duralumin pipes.

The exhibited "Albatross-20" deltaplane was given to the AF Museum on March 10, 1983, by students of the Moscow Higher Technical School, named after N. Bauman.

# 10

## Lighter-Than-Air Aircraft

## "The USSR-1" Stratospheric Balloon Car

The first Soviet "USSR-1" stratospheric balloon car was created in 1930 by a group of designers under V. Chizhevsky.

The stratospheric balloon is designed for scientific investigations of the upper atmosphere layers, the study of cosmic ray intensity, and penetration capability of air composition and pressure at various altitudes.

The stratospheric balloon car volume - 6.2 m cubed, mass - 280 kg, ballast mass - 620 kg.

The car is made of chain-armored aluminum, it has 2 access hatches with a diameter of 500 mm, 7 windows with a diameter of 100 mm, osier shock-absorber, and air-navigation instruments for scientific research.

To maintain normal temperatures inside the car it is covered with light deer felt and by linen skin.

A normal temperature of 20-30 degrees C was maintained inside the car.

On the 30th of September 1933 "The-USSR-1" stratospheric balloon with a crew on board (commander G. Prokofiev, radioman E. Virnbauman and engineer K. Godunov) performed a flight to an altitude of 19,000 meters, breaking the world altitude record. The car took off at 8:40 a.m. from the Frunze Central Moscow Airfield and landed at 17:00 hours in Kolomna.

The exhibited "USSR-1" stratospheric balloon car was delivered to the AF Museum from the Central House of Aviation and Cosmonautics named after meters. Frunze in 1960.

# "The Volga" Stratospheric Balloon Car

"The Volga" high-altitude stratospheric balloon-laboratory was designed for conducting medical-biological research, and also for the investigation of high-altitude vehicles' crew recovery. It was built in 1962.

The stratospheric balloon characteristics: flight altitude - up to 25,100 meters; take off mass - 27,000 kg; the balloon envelope volume - 72,000 m cubed. Time to climb to 25,000 is normally about 3.5 hours, and when depressurized - about 2 hours. The parasuspending system for descending after the balloon envelope release consists of a stabilizing canopy with an area of 30 square meters, and the main canopy with a 1,400 square m area.

The car structure and equipment includes: life-supporting system; the balloon flight control system parachute and the balloon envelope release control system; electric equipment and power sources; instrumentation; measurement and recording equipment; and a shock-absorbing device. The crew is 2 men.

The "Volga" stratospheric balloon took off to an altitude of 25,458 m in November 1962.

The crew, consisting of Colonel Pyotr Dolgov—commander of the balloon—and Major Eugene Andreev conducted a number of scientific investigations, tested the new aeronautical and space equipment recovery means, and performed record jumps.

E. Andreev ejected and after 270 rounds of free fall he was at 24,500 meters, having opened the parachute at 958 meters.

P. Dolgov left the car through the egress hatch in 1 minute 30 seconds after E. Andreev's ejection. However, on leaving the car his suit was depressurized and P. Dolgov perished.

The exhibited "Volga" stratospheric balloon car was delivered to the AF Museum on February 27, 1964.

# Spherical Balloons Baskets

Spherical balloons were used for training aeronauts, for sport flights, and for various scientific research.

This spherical balloon has the following main parts: the balloon envelope inflated with gas, the heat suspension system, and a basket.

The basket accommodates the crew, the ballast (sand or shotbag ballast)—a proportioned discharge which is used for gaining climb speed or descent—food, and instruments assembled on a frame or in a special case. The set of instruments in the basket depended on the flight program: barograph, altimeter, clock, variometer, thermometer, compass, psychrometer, air sextant, binoculars, the chart board, first-aid kit, and so on.

The whole crew had parachutes. The basket had elementary devices for the crew rest, such as seats and a swinging down deckchair. In case of a scientific research flight the basket accommodated special equipment.

The oxygen equipment was switched on at an altitude of 4,500-7,000 meters.

The exhibited balloon baskets were delivered to the AF Museum from the Frunze Central House of Aviation and Cosmonautics on November 23, 1987. The basket accommodates special equipment.

# 11

## Aircraft Armament

## Armament

### PV-1 Machine Gun

It was developed in the late 1920s under the guidance of A. Nadashkevich and represents the re-design of the famous infantry "Maxim" machine gun.

A. Nadashkevich reduced the mass of the machine gun from 20 to 14.5 kg and increased its rate of fire from 600 to 780 rounds per minute. High-rate firing was used only when it was necessary. The machine gun caliber is 7.62 mm. The round mass - 22.4 g. The bullet mass - 9.6 g. The cartridge-case mass - 9.6 g. The machine gun length - 1,050 mm. The initial bullet velocity - 870 m/sec. The average barrel pressure - 2,750 atm. The bullet kill capability from 200 steps distance is up to 40 planks 2.5 centimeter thick. It has the same beltfeed as the "Maxim." There is a re-arming stick for the machine gun re-arming in the air and also on the ground for eliminating some delays in the air.

The PV-1 machine gun was installed in I-2, I-3, I-5, I-15, and I-15 bis fighters for synchronized firing though the aircraft screw from a fixed position, and could not solve the problem of the aircraft armament completely.

The exhibited machine gun was delivered to the AF Museum from a small arms dump in February, 1959.

### DA Machine Gun

Infantry hand machine gun originally designed for a 7.62 mm caliber rifle cartridge created by V. Degtyaryov was introduced into the Red Army service in 1927. Earlier Degtyaryov, together with V. Phoyodorov, developed a special 6-mm caliber air machine gun on the basis of an infantry submachine gun, which naturally required a special cartridge.

Reconnaissance and bomber aircraft needed turret air machine guns. For this purpose Degtyaryov redesigned the gun so it could be used by aircraft.

The small mass of the machine gun was achieved due to the right choice of the automatics work principle based on the elbowing of a part of the gunpowder gases into a special gas camera.

All the parts of the mechanisms were arranged on the bolt frame, which facilitated stripping and assembling the machine gun.

The designer introduced the following changes into the air machine gun system: the jacket of the machine gun was removed to decrease the machine gun mass and dimensions, and also for the better cooling of the barrel by air flow; an air sight was mounted on the front part of the barrel bon; the muff for weathercock front-sight fastening was on the muzzle side of the barrel; the machine gun, through special devices, was connected with the turret mount of the aircraft; the butt was replaced by two hilts for comfortable firing; the single row disk ammunition drum was replaced by a three-row magazine of larger storage capacity (for 65 cartridges) and smaller diameter; and the case collector box was introduced.

In 1928 the Degtyaryov system (DA) turret 7.62 mm machine gun was introduced into the inventory of Soviet AF units as a defensive weapon for bomber and reconnaissance aviation.

The machine gun mass - 7.1 kg. The bullet mass - 9 g. The cartridge mass - 23 g. Rate of fire - 500-600 rounds per second. The bullet initial flight velocity - 826 m/s. Because of the low rate of fire twin mounted DA-2 machine guns were used on the turret. So, TB-1, the first Soviet bomber, was armed with three turret twin mounted DA-2 machine guns, and TB-3 had eight DA and DA-2 machine guns. The exhibited DA machine gun was received from the ammunition depot in February, 1959.

# ShKAS Machine Gun

Fit to aircraft, infantry machine guns could not satisfy the needs of aviation completely. Aircraft armament must be more effective, be of a higher fire rate, have less kick, a belt feed, and fire and re-arming must be by remote control. That is why many design offices worked on the problem of special aircraft weapons development.

The "ShKAS" machine gun, developed by the Design Office headed by B. SHPitalny and I. Komarnitsky, had a record-breaking fire rate of the time (up to 1800 rounds per minute). It was introduced into service in 1932.

In 1932 the development of another ShKAS variant was successfully completed by K. Rudnev in cooperation with V. Kotov, V. Salishchev, and V. Galkin.

"ShKAS" was mounted on many types of aircraft up to the Great Patriotic War (WW II). The cartridges for "ShKAS" were done under the guidance of N. Yelizarov and had tracer, incendiary, combined action bullets, and armor-piercing incendiary bullets capable of igniting armored fuel tanks. These raised the efficiency of the machine gun to a considerable degree.

Later the designers of the "ShKAS" machine gun created the "Ultra-ShKAS", with a 4,000 rounds per minute rate of fire, and the armorers from Tula constructed a mechanical paired "ShKAS" machine gun, having increased the fire rate up to 6,000 rounds per minute.

# SN Machine Gun

In the 1930s designers I. Savin and A. Norov developed an air gun named SN. When its barrel passage was released, the shutter and the barrel shifted to the contrary sides and then met when locked. Due to this invention movable parts were reduced and the rate of fire became 3,000 rounds per second. However, the machine gun did not go into service and became the last designed for a rifle 7.62 mm cartridge.

The 1930s were characterized by the rapid development of aviation. Flight speeds and aircraft survivability increased. The machine guns of "normal" caliber became ineffective. They were replaced by cannons and large-caliber machine guns.

# A-12.7 Machine Gun

Developed by N. Afanasiev, it was the first Soviet large-caliber machine gun designed for helicopters.

It was introduced into inventory in 1952. Used in the A-12.7, the original lay-out of automatics with the booster which delivered the cartridge from the link of the ammunition belt to the barrel opened an important stage in the development of the gas-operated armament.

Its mass is 23 kg. The projectile mass - 40-50 gr. Caliber - 12.7 mm. Rate of fire - 800-1000 rounds per minute. Salvo weight per second - 0.52-0.92 kg. The initial bullet flight velocity - 785-820 m/s.

# UB Machine Gun

In 1939 meters. Berezin, a designer from Tula, together with Z. Mamontova and A. Chepelcv developed a large-caliber UB machine gun which could be used in wing-mounted (UBV), turret (UBT), and synchronized variants. The UB abbreviation means "Universal, of Berezin."

The machine gun mass - 21.5 kg, caliber - 12.7 mm. The bullet mass - 48 gr. Rate of fire - 1,000 rounds per minute. Initial bullet flight velocity - 860 meters per second. Salvo weight per second - 0.8 kg.

That machine gun was introduced into the inventory in 1940. During the war it became one of the main types of aircraft armament. In 1941 the Izhevsk plant produced 6,300 UB-12.7mm machine guns. In 1942 the plant produced 25,000 of them, and in 1943 more than 40,000, which met the requirements of aviation industry.

The machine gun replaced the "ShKAS" machine gun successfully.

This machine gun was installed in "Yak-1", "Yak-7", "Yak-3", "MiG-3", "Pe-2", "Il-2", "Tu-2", "Pe-8", and "Il-4" aircraft.

The exhibited UB machine gun was delivered to the AF Museum in 1958.

# ShVak Cannon

The large-caliber machine gun designed by S. Vladimirov on the basis of the "ShKAS" in 1934 had become an intermediate between the high fire rate "ShKAS" machine gun and the aerial "ShVAK" cannon.

The machine gun survivability was enough to allow S. Vladimirov (V. Degtyaryev Design Office) in 1936 to make the bicaliber machine gun and to create the 20 mm "ShVAK" cannon.

A new unit of the structure was the two-cycle belt-feed—the ammunition belt moved when the traveling unit recoiled and recuperated.

First the cartridges with fragmentation and tracer shells were used. Later, armor-piercing-incendiary and fragmentation-incendiary projectiles were developed.

The cannon caliber is 20 mm. The cannon mass - 42 kg. Rate of fire - 800 rpm. The projectile initial velocity is 800 m/s. The projectile mass - 96 gr. Salvo weight per second - 1.28 kg.

Pressing work carried out by the Polikarpov Design Office resulted in the I-16 being equipped with two "ShVAKs" in addition to two "ShKASs"

"ShVAK" cannons were used by I-16 fighters in the Khalhkin-Gol hostilities in 1939. The "ShVAK" cannon was used both in synchronous and wing variants, and was mounted on A. Yakovlev and S. Lavochkin fighters, S. Ilyushin attack aircraft, and the V. Petlyakov long-range bomber.

# VYa Cannon

In 1940 A. Volkov and S. Yariyev designed an aerial VYa cannon of 23 mm caliber. Its principle of operation was based on elbowing gunpowder gases through a hole in the barrel channel. Armor-piercing-incendiary and fragmentation-incendiary projectiles with quick fuses were developed for this cannon.

There were no cannons of such caliber abroad at that time.

In May 1941 the cannon was adopted for service and produced in large numbers by several plants.

The cannon was in service with fighter and attack aviation.

The "VYa" cannon was designed for operation out of propeller planes (wing variant) or through the motor reduction gearbox bush (motor variant).

The projectile mass - 197 gr, rate of fire - 550-650 rpm, the projectile initial velocity - 900 m/s. The cannon mass - 66 kg. Salvo weight per second - 2 kg.

High projectile initial velocity, together with a powerful destructive effect, allowed the "VYa" cannon to be used successfully, not only against aerial targets, but ground targets as well.

When the projectile hit a fuel tank, it exploded. If it was a wing, there appeared a shot-hole up to 0.75 square meters in size. The projectile of the "VYa" cannon pierced armor 25 mm thick from a distance of 400 meters.

After each cannon burst the traveling units stopped at the rear in the extreme rear position, which excluded the possibility of the projectile ignition in the cartridge chamber getting warm during firing. The cannon is recharged pneumatically. The pressure of the compressed air in the pneumatic recharging cylinder is 30-35 atm.

# B-20 Cannon

In 1941 on the basis of the UB-12.7 mm machine gun Berezin developed a 20 mm B-20 cannon for "ShVAK" gun cartridges.

In October 1944 the cannon was introduced into service.

The B-20 cannon was designed in three variants: synchronized, turret, and wing-mounted.

The cannon mass - 25 kg. The projectile mass - 96 gr. Salvo weight per second - 1,28 kg. Rate of fire - 800 rpm. The projectile initial velocity - 800 m/s.

The B-20 has practically the same ballistic characteristics and the rate of fire as a "ShVAK" gun, but is almost two times lighter.

The B-20 cannon was fixed mounted on the fighters "Yak-3", "Yak-7B", "LaGG-3", "La-5", and "La-7." On "Il-2" attack aircraft and "Tu-2" and "Tu-4" bombers it was of a flexible type.

# NS-37 Cannon

In 1941 designers A. Nudelman, A. Suranov, G. Zhyrnykh, V. Nemenov, S. Lunin, and meters. Bundin (under the guidance of A. Nudelman and A. Suranov) developed a 37 mm NS-37 cannon.

That cannon had a short barrel stroke, spring-hydraulic recuperator, and cartridge belt-feed. The belt-feed mechanism was activated by a special spring raising when the barrel recoiled.

In 1942 the NS-37 cannon was accepted for service and put into mass production. 40 cannons were produced in 1942, and 4,730 in 1943.

Its fire unit consisted of armor-piercing-incendiary-tracer and fragmentation-incendiary-tracer projectiles. Armor-piercing charges were designed for destruction of ground armor and fragmentation charges were designed for aerial targets.

One NS-37 cannon projectile hit was enough to destroy an aircraft. The armor-piercing projectile pierced 50 mm thick armor from a distance of 200 meters.

The cannon mass - 150 kg. The projectile mass - 735 gr. Rate of fire - 250 rpm. The projectile initial velocity is 900 m/s. Salvo weight per second - 3-6 kg.

The NS-37 cannon was mounted on "LaGG-3", "Yak-9T" (in the blocks chamber), and on "Il-2" (under the wings).

# NS-23 Cannon

In 1944 the A. Nudelman Design Office developed a new NS-23 light aircraft cannon of 23 mm caliber with reduced projectile initial velocity. The masses of the cannon and its dimensions were considerably smaller than those of the "VYa" cannon.

The "NS-23" is of an analogous structure to the "NS-37". Small mass and dimensions of the "NS-23" cannon and a small kick in firing considerably simplified its installation and operation.

A special cartridge and link were designed for the "NS-23." They had a mass two times smaller than that of the "VYa" cannon, allowing for an increase in the ammunition. In October 1944 the "NS-23" cannon and its cartridge were introduced into service and serial production.

The cannon mass - 37 kg, the projectile mass - 200 gr, rate of fire - 550 rounds per minute. Salvo weight per second - 1.85 kg; the projectile initial velocity - 690 m/s.

The "NS-23" cannon was mounted on "Il-10", "La-7", "La-9", "La-11", and "MiG-15" aircraft.

# N-37 Cannon

The successful creation of a lighter cartridge and the "NS-23" cannon proved the work on the cartridge and "NS-37" cannon to be expedient. Decreasing the larger power, the designers improved the cannon automatics cycle structure, making its rate of fire almost twice as high and decreasing its weight and dimensions. The shot power influence on the aircraft was reduced. The first prototype of the "N-37" cannon was manufactured in April 1944. The cannon's first air firings were conducted from Yak-9 aircraft.

The cannon mass - 10.3 kg. The projectile mass - 735 gr. Rate of fire - 400 rpm. Salvo weight per second - 4.9 kg. The projectile initial velocity - 690 m/s.

# NR-30 Cannon

When the designers began to mount the cannon on jet aircraft they faced a new problem: firing from the cannons, accommodated in the aircraft nose, made the engines fail due to gas penetration into the suction nozzle. The problem was solved structurally, and, beginning in 1947, the "N-37" cannon became the main armament of jet aircraft, such as the "MiG-9", "MiG-15", "MiG-15 bis", MiG-17", and "Yak-25."

The jet fighters' performance increase in the 1950s demanded the development of corresponding high fire-rate weapons.

Several design offices worked on the 30 mm cannon. The "NR-30" cannon designed by A. Nudelman and A. Rikhter turned out to be the best one, and it was introduced into the inventory in 1955. It differed from the previous prototypes by a number of specific solutions: the spring-hydraulic recuperator was replaced by a gas one, and the cartridge-feed mechanism was made two-stroke and two-way. The cannon ammunition and the cannon itself were developed simultaneously. The "NR-30" became the main weapon of Soviet fighters.

The cannon mass - 66 kg. The projectile mass - 410 gr. Rate of fire - 900 rpm. Salvo weight per second - 6.15 kg. The projectile initial velocity - 780 m/s.

# GSh-23 Cannon

The potential of one barrel aerial cannon fire rates having been exhausted led to the advent of multi-barrel cannons.

The two-barrel automatic "GSh-23" cannon designed by V. Gryazev and A. Shypunov was introduced into service in 1965.

The automatic operation is based on the employment of gun-powder gases elbowed from the barrel channel. Firing from the barrels is performed in turn due to kinematic coupling between the right and the left barrels.

It had a belt feed, and the cannon recharging was carried out with the help of a special explosive charge.

The cannon caliber -23 mm. Its mass - 51.5 kg. Rate of fire - 3,000/3,400 rpm. Fail-safe of each barrel is 4,000 rounds. The projectile initial velocity - 715 m/s.

# GSh-6-30 Cannon

This is an aerial six-barrel rotary cannon. The automatic operation is based on employment of gun-powder gases elbowed from the barrel channel as in the case of the GSh-23 cannon.

The barrel unit rotation is performed with the help of a crank-connecting rod mechanism.

As with the GSh-23, the cannon has the belt feed in the same part. The cannon cartridges have an electric primer.

The cannon caliber - 30 mm. Rate of fire - 5,000-6,000 rpm.

The cannon mass - 140 kg. The projectile initial velocity - 900 m/s. Fail-safe of each barrel is 6,000 rounds.

# Bombs

## OFAB-100M Bomb

The AF Museum exhibits a great number of aerial bombs, all much alike in design but of different functions and caliber.

Among them there are "OFAB-100-120", "OFAB-250-270", "OFAB-250 ShH", "OFAB-250 (U-1350)", and "OFAB-100M" general purpose bombs. Let's examine one of them.

The "OFAB-100M" general purpose aerial bomb is designed for destroying personnel, tanks, armored personnel carriers, and other material. It consists of a body, ballistic ring, stabilizer, and an explosive substance. The body is welded, and consists of a warhead, cylinder, and the tail cone. The warhead is an oval punched out of sheet steel. The warhead fuse barrel bush is threaded and designed for a fuse 36 mm in diameter, or for screwing an adapting bush with a bomb fuse cavity of 26 mm.

The fragmentation section of the bomb is the body cylinder, which contains the main mass of metal (the cylinder webs are 27 mm thick).

The tail cone of the body is rolled of sheet steel and welded in the main seam. The stabilizer is of a box-type and consists of four fins. The tail bush has a threaded bomb fuse cavity of 36 mm.

A ring is welded to the bomb cylinder part to load aircraft bombs.

The aerial bomb weight (without the fuse) is 121 kg. The explosive substance weight - 36 kg. The radius of all-round destruction zone is 36 meters.

The bomb is outfitted with the K-2 mixture, or 90 percent trotyl and 10 percent dinitronaphthalene mixture. The "OFAB-100M" bomb was delivered to the AF Museum on April 12, 1964.

## AO-10SCh Bomb

The "AO-10SCh" aerial fragmentation bomb is designed for destruction of light armor protection targets and the personnel following them.

It consists of a body, adapting bush, stabilizer, and explosive substance charge. Its body tail, welded of steel cast iron, is filled with standard threaded stud designed for stabilizer fastening.

A "AM-L" threaded fuse bush is screwed into the warhead.

The "AO-10SCh" bomb stabilizer is box-shaped. The fins are tied by electric welding.

A concave pad with a hole in it is welded to the upper part of the fins; with the help of this pad the stabilizer is mounted onto the bomb body and fastened to the stud with a nut. The AO-10SCh aerial bomb is outfitted with K-2 mixture.

The exhibited "AO-10SCh" aerial bomb was delivered to the AF Museum on October 2, 1961.

# FOTAB-100-140 Bomb

"FOTAB-100-140" photoflash aerial bomb is designed for night aerial photography from an altitude of 15,000 m and at aircraft speeds of 2,500 kph.

The "FOTAB-100-140" bomb consists of a steel body, stabilizer, and loading. The body includes three sections: massive warhead, cylinder, and tail cone.

The steel warhead is oval. The warhead bushing has a bomb fuse cavity of 26 mm diameter. The cylinder section of the body is used for accommodation of the photoflash and ignition explosive charge and is a small, web steel cylinder. A pad of sheet steel is welded to the outer side of the cylinder.

The tail cone is made of sheet steel and welded with the cylinder.

A part of the bomb loading is accommodated in the tail cone.

The bomb loading consists of photoflash compound, ignition-explosive charge, seven detonating grains, and a gear ration charge.

The bomb length - 1,465-1,478 mm, the body diameter - 240 mm. Its mass - 137 kg, the photoflash compound mass - 51.5 kg. The ignition—explosive compound mass - 14.3 kg.

The "FOTAB-100-140" aerial bomb was delivered to the AF Museum on August 23, 1983.

# SAB-3M Bomb

The "SAB-3M" is a flare bomb designed for illuminating the terrain and the objective's location to ensure visual reconnaissance and pinpoint bombing at night.

The aerial bomb consists of the warhead cylinder section and base cap. Inside the body there is a flame (torch) obturator and parachute. The parachute is rolled up into a cylinder and packed up into the body; the flame is separated from the parachute by the obturator, consisting of a felt ring and two metal disks fastened with four rivets.

The "SAB-3M" was received by the AF Museum from the AF Engineering Academy, named after Professor N. Zhukovsky, on May 22, 1959.

# ZAB-50TG

The "ZAB-50TG"—an incendiary aerial bomb with solid fuel, designed for destruction of buildings, depots, and other targets.

The "ZAB-50TG" consists of the warhead cylinder, hard ring, connecting ring, rear section bushing, base lock, lug, stabilizer, and loading.

The bomb mass - 48 kg, the explosive substance mass - 26 kg. The filling factor - 0.59. The body diameter - 0.203 meters, the length - 0.99 meters. The flame length is approximately 4-5 meters. The time of burning is about 5-6 minutes.

The incendiary aerial bomb with solid fuel has a rifled bomb fuse cavity and four gas-operated holes sealed with ceresine.

The "ZAB" is equipped with a direct impact quick fuse with the primer-igniter.

A box-shaped stabilizer is welded onto the tail-group of the bomb.

The "ZAB-50TG" was delivered to the AF Museum on May 22, 1959.

## FAB-High-Explosive Aerial Bomb

High-explosive (HE) aerial bombs are designed for destruction of buildings, railway junctions, ships, bridges, and other military objectives by gas set free in the explosion. The HE bomb consists of a body and a stabilizer.

Its body is designed for filling the bomb with explosive substance. In turn the body includes a warhead cylinder and tail unit. The warhead is oval and is made of steel and cast iron. The cylinder section is made of a sheet of steel 4-20 mm thick (it depends on the caliber) rolled up into a cylinder and fastened by welded joints.

The tail unit is produced in the form of a carved cone. It improves the bomb flow, accommodates the fuse, and fastens the stabilizer.

The AF Museum exhibits HE bombs of different caliber from the smallest to the largest (FAB-5000 and FAB-9000).

## PTAB-2.5-1.5 Anti-tank Aerial Bomb

In 1942 I. Larionov proposed a design of a light anti-tank aerial bomb. The bomb tests were finished on April 21, 1943. It could penetrate armor 70 mm thick, and was so effective that the USSR State Committee of Defense urgently decided to introduce "PTAB-2.5-1.5" into service and to organize its mass production.

The Peoples' Committee of Ammunition was assigned the task of manufacturing 800,000 "PTAB-2.5-1.5" bombs with ADA fuses by May 15, 1943.

The "PTAB-2.5-1.5" combat employment demonstrated its high efficiency against the main force of German Wehrmacht tanks and self-propelled guns.

"PTAB-2.5-1.5" consists of a warhead, body, tail unit, and shaped charge.

The bomb mass - 1.6 kg, length - 360 mm. Diameter - 66 mm. The explosive substance mass - 0.610 kg. The explosive substance compounds - 90 percent of trotyl and 5 percent of aluminum powder. The kill radius of personnel by shell - splinters in 5-10 metres.

# Missiles

## RC-2US Missile

The "RS-2US" air-to-air fragmentation high-explosive missile is designed for destruction of air targets by "MiG-21PFM", "Su-9", and "MiG-19PM." Missile launch may be individual or in salvo.

The missiles are on underwing stores with special launching devices and after the launch they are guided by radio signals of aircraft intercepts and sighting stations.

The "RS-2US" missile is of a canard configuration with horizontal surfaces in front of the wings.

The missile equipment ensures its guidance to the target along the equisignal line of the aircraft radar beam by the "three point" method.

The missile consists of a body with wings and control surfaces; combat loading; gun-powder rocket engine; electrical system; and air system.

The missile mass is 83.2 kg. The warhead mass - 13 kg. The guided flight time - 12 seconds. The employment temperature - +60-50 degrees C. The employment height is 500 m up to the aircraft service ceiling. The missile is used against the targets flying at speeds exceeding 650 kph with an aspect angle of 0-2/8. Its launching distance at the altitude of 4,000 m is 2-7 km and at the lower altitude - 1.5-5.5 km.

(See photo on page 168, center)

# R-3-S

The "R-3-S" is an air-to-air homing missile with heat-seeking coordinator. It is designed against air targets under simple weather conditions in daytime and at night. The missile homing is performed by the method of proportional close-in.

When the target is within the optical sight of the infra-red homing head the pilot hears an audible signal, which means the missile is homing on to the target and sighting accurately.

At the launch moment the missile coordinator locks the target and automatic tracking begins.

After the launch the missile flies along the straight line (with a zero signal on the control unit) for some time, and then it performs a maneuver towards the range line (missile-target line) to take the necessary lead angle.

The independent flight and transition to the parallel closing with the target takes 3-5 seconds, which depends on the initial conditions of the launch.

The "R-3-S" missile is of a canard configuration, with X-mode control surfaces and wings.

The missile mass - 75.3 kg. The warhead mass - 11.3 kg. The missile length - 2,838 mm. The wing span - 528 mm. The employment temperature - +50 degrees C.

The firing range - 1 - 7.6 km. The firing interval 3-4 seconds. The launch aspect angle - 0-6/8. Minimum firing range - 0.6-0.8 km. The guided flight time - 21 sec. Operational height - 0-21.5 km. Self-destruction time from the launch moment - 21-26 seconds.

The minimum G-load at the launch time - 2. Maximum fuse radius - 9m.

(See photo on page 168, right)

# R-60 Missile

The "R-60" air-to-air all angle homing missile with heat-seeking coordination is designed for the destruction of enemy aircraft in daytime and at night under instrument weather conditions in close air combat.

The missile mass - 45 kg. The warhead mass - 2.7 kg. The missile range is 900-1,200 m at low altitudes, and 15-20 km at high altitudes.

The missile length - 2,095 mm. Its diameter is 120 mm. Span - 390 mm. Operation time of the gun powder engine is 3-5 seconds; in 23-29 seconds after launch the missile is self-destructed.

The "S-5M" missiles can also be used for destruction of ground targets (aircraft in parking places, automobile fuel dumps, etc.).

The firing is performed by special single-charge guns with the open breech accommodated in UB-8, UB-16, and UB-32 units.

The missile includes the body, the warhead, and gunpowder jet engine.

The missile mass - 3.85 kg, caliber - 57 mm. The length - 882 mm. The minimum firing range - 0.6 km, maximum - 3 km. The destruction height - 0.2-10 km. The firing may be performed in salvos of 4, 8, or 16 missiles by two units simultaneously. The maximum missile flight speed - 725-665 m/s.

## S-21 Missile

The "S-21", a jet fragmentation high-explosive projectile with an electric fuse, is designed for launch by fighters and fighter-bombers against enemy combat formations, individual bombers, and also against ground targets with light armor protection or without it (aircraft on parking crossings, etc.)

The firing is performed by "Pu-21" aircraft from built-in launchers.

The "S-21" missile consists of a body, a warhead, a gunpowder jet engine, and a stabilizer.

The projectile caliber is 212 mm, and its length with the fuse - 1,756-1,757 mm.

The projectile mass - 117.8 kg. The explosive charge mass - 10.6 kg. The missile flight time over the active leg - 0.79-1.13 sec.

## S-5M Missile

The "S-5M" unguided jet high-explosive projectile with a mechanical fuse of the shock type is designed for the destruction of enemy bombers at an altitude of 30,000 meters. Its firing is performed by fighters, fighter-bombers, and attack helicopters.

## S-24 Missile

The "S-24", a jet fragmentation high-explosive missile, is designed for launching by fighters and fighter-bombers against ground targets, artillery batteries, missile launching sites, enemy personnel and material, crossings, trains and railway stations, ships, etc.

The projectile is fitted with a power-driven V-24A contact and time delay fuse.

The missile consists of a warhead, body, and a gunpowder jet engine with stabilizer.

The projectile caliber - 240 mm, length - 2,220 mm, mass - 235 kg. The minimum operational range is 1.6 km, maximum - 3 km. The minimum altitude of operation - 0.5 km, maximum 4-5 km. The maximum projectile flight speed - 410 m/s. The missile pierces the armor 25 mm thick. Two rounds salvo has an interval of 0.23 seconds.

# S-3K Missile

The S-3K-air jet projectile of the cumulative type is designed for firing by fighter-bombers against ground small-sized armored targets, such as tanks and self-propelled artillery pieces. "S-3K" projectiles are fired by multi-loading APU-14U launchers.

The S-3K consists of the body, warhead, powder jet engine, and EVU-84 electric fuse.

The projectile mass with the electric fuse - 23.5 kg. Its length - 1,500 mm. Maximum projectile flight velocity - 360-380 m/s. The projectile flight time is 1,000 m for 3.82-3.62 seconds.

The ruptured armor - 300 mm.

# KS Winged Projectile

The KS winged projectile is an unmanned flying vehicle designed for destruction of enemy surface ships. It resembles a MiG-15 aircraft of the same design office, but half-sized.

The flight mass of "KS" - 2,735 kg. The wing span - 4.7 meters. Sweepback of leading edge - 57.5 degrees. The fuselage length - 8.3 meters. It has a modified V. Klimov RD-500K engine.

The projectile is launched by "Tu-4" carrier-aircraft. The missile guidance is based on the principle of semi-active radar. The transmitter is mounted on the carrier-aircraft, and the receiver is on the projectile. After the release, the "KS" automatically follows the radio beam emitted by the carrier plane and at the flight terminal—the target echo.

The "KS" development was performed on the manned analogue. The flights were performed by test-pilots S. Amet-Khan, S. Anokhin, Ph. Burtsev, V. Pavlov, and P. Kazmin. At that stage of development the "KS" projectile had a bicycle landing gear. The analogue performed 150 flights.

In its main unmanned variant the projectile does not have a cockpit, manual control system, or landing gear. The military load has 500 kg of trotyl.

Out of 13 test flights, 9, including the last one on September 21, 1952, carried out with military loading, resulted in a direct hit of the target.

# Kh-66 Missile

The "Kh-66" is an air-to-surface fragmentation and high-explosive guided missile. It is designed for destruction of small-sized ground and sea-surface targets. The target detection is visual. The missile is guided onto the target by a radar sight beam. The TV guided missile flies to the target following the beam of equisignal direction, emitted by the aircraft radar sight antenna.

The "Kh-66" missile consists of a warhead with fuse, control surface, gunpowder rocket engine, roll stabilization, and control units.

The maximum firing range is 8-10 km, optimum range - 6 km. The minimum range - 3 km. Launching speed - 650-850 kph. Launching height - 5,000 meters (optimum - 3,000-5000 meters). The minimum guided flight time - 20 seconds. The missile may be launched individually or in salvo.

The missile mass - 277.8 kg. The warhead mass - 105+4 kg. The split number exceeds 2,000. A split mass - 14.5 kg.

A split initial speed - 2.300 m/s. The gun-powder rocket engine work time is 4-6.7 sec. The missile length - 3.631 meters. Span - 0.275 meters. The power plant total thrust - 10,850 kg HP.
(See photo on page 168, left)

# Kh-23 Missile

The "Kh-23" is an air-to-surface fragmentation and high-explosive CP and TV guided missile. It is designed for destruction of small-sized ground and sea-surface targets. The target is detected visually. The carrier-aircraft is fitted with "Delta" guidance equipment. The missile guidance may be performed automatically or manually.

Guiding the missile manually, the pilot corrects the missile flight with the help of an erection knob, and in automatic flight the missile is erected by the automatic tracking system.

The missile consists of control surfaces, warhead, gunpowder rocket engine, and control units.

The missile mass - 298 kg. The warhead mass - 111 kg. The missile length - 3,591 mm. diameter - 275 mm, span - 788 mm. The firing range - 3-10 km. The destruction height - 0.1-10 km.

# 12

## Recovery Aids

### MiG-19 Ejection Seat

The first recovery ejection installations were developed during World War II when high flight speeds made the aircraft crew's escape very difficult, and at speeds exceeding 400 kph, almost impossible. The first ejection seats appeared in Germany, then in Great Britain.

The first experimental pilot ejection from a "Tu-2" in the USSR was performed by G. Kondrashev on June 24, 1947.

Later, the majority of the home design agencies developing combat jet aircraft foresaw the installation of ejection seats of Soviet-made design.

A typical representative of that development is exhibited in the AF Museum ejection seat of a "MiG-19" aircraft, ensuring emergency escape at a speed of 900 kph with a 20 G load. The minimum ejection level in horizontal flight is 200 meters, in gliding at the descent speed of 20-40 m/s - 350-400 meters; in diving with the 30 degree dive angle - 900 meters, in 60 degree angle dive - 1,400 meters. So, the pilots' rescue in emergency situations was not guaranteed in all flight envelopes.

The pilot's seat belts were automatically unfastened by an AD-3 device in 1.5 seconds after the ejection.

After the ejection seat escape, the pilot's parachute opened automatically at a level of 1,000 m over the terrain, or by a KAP-3 device at a lower level in 2 seconds.

### KS-4-22 Ejection Seat

The "KS-4-22" is the result of recovery aids unification work and expansion of their operational employment.

The "KS-4-22" is an ejection seat designed for a pilot's escape at all altitudes from take-off to landing at instrument speeds from 140 kph to 1,200 kph with the G-load of 20.

The ejection seat includes the following components:

- KSM—combined gun mechanism consisting of two stages: the 1st stage—gun powder ejection device, the 2nd - two chambers of gun powder rocket engine with response time of 0.4 sec developing thrust of about 5 tons;

- PS—S three-cluster parachute system;

- KAP-4, PPK-I—P devices ensuring automatic activation of all elements of the system;

- KAP-4—automatic ejection device: its response time at the speed of 500 kph is immediate, and with a delay of up to 1.6 sec at speeds of 540-1,200 kph;

- PPK-I—P is a combined parachute device with fast response at low levels and with a fixed parachute delay at medium and high altitude;

- NAZ—7-life supporting device after water-landing. This installation ensured the pilot's recovery at almost all speeds and altitudes in the range of the second generation jet aircraft.

# K-36DM Unified Ejection Installation

The "K-36DM"—a unified ejection installation designed at "Zvezda" scientific (industrial association under Chief Designer G. Severin) is used by all types of military aircraft of the 1970s-1980s and later. It provides a reliable escape in practically all flight speed and altitude ranges, including take-off and landing runs. The G-load in ejecting reaches 20 units of measure.

The ejection installation includes the following main components:

- seat (seat pan, back head-rest);
- energy source (two-stage ejection mechanism, pyrotechnical cartridges, pilot fixing, head rest unit and pilot seats);
- wind-blast protection unit (deepened head-rest, arms spread stop, legs scatter stop with jack, deflector);
- stabilizing unit (two stabilizing parachutes);
- PL - 72 recovery parachute;
- ejection control system consisting of two sticks, interlock mechanism preventing ejection with unjettisoned canopy and canopy ejector;

- oxygen supply ejector;
- articles of life—support system: NAZ-8, PLOT PS N-1, Mosquito (Komar)—2M automatic radio beacon, signal means.

To perform ejection the pilot is to meet only two requirements: grip the stick and pull it out—the rest will be done automatically.

The experience of the ejection seat operation proved its extraordinary reliability.

Without any exceptions the second series "K-36DM" emergency employment by "MiG-29" and "Su-27" aircraft resulted in the pilots' uninjured recoveries. There was even involuntary static ejection, which turned to be successful.

In 1989 at Le Bourget (Paris), during a demonstration flight, the "MiG-29" engine failed at 92 m altitude, and the famed test pilot Anatoly Kvotchur ejected. The ejection seat operated reliably and the pilot was not injured.
(See photo on page 174)

*Also from the publisher*

## Protect & Avenge:
### The 49th Fighter Group in World War II

S.W. Ferguson & William K. Pascalis

The World War II-year's odyssey of the most successful fighter group in the war against Japan.
Size: 8 1/2" x 11" over 600 photographs, color aircraft profiles, maps
360 pages, hard cover
ISBN: 0-88740-750-0          $49.95

## McDonnell-Douglas
## F-15Eagle:
### A Photo Chronicle

Bill Holder & Mike Wallace

Photo chronicle covers the F-15 Eagle from planning and development to its success in Operation Desert Storm and post-Desert Storm. All types are covered, including foreign—Israel, Japan and Saudi Arabia.
Size: 8 1/2" x 11"    over 150 color & b/w photographs
88 pages, soft cover
ISBN: 0-88740-662-9          $19.95

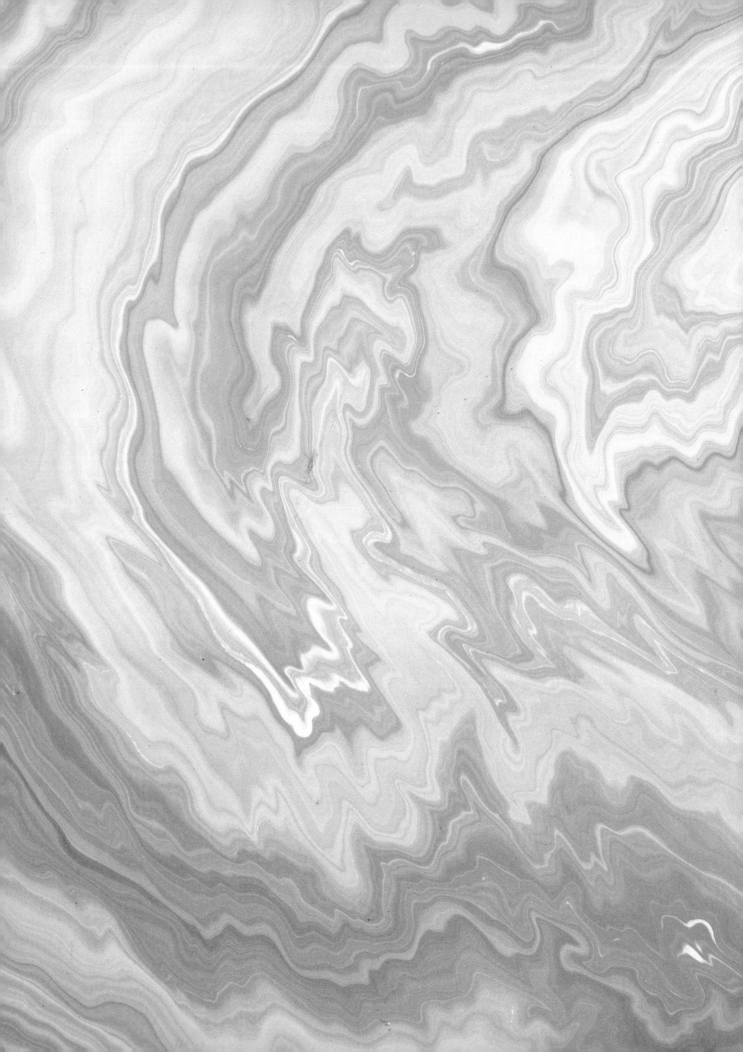